Heat-integrated exhaust purification for natural gas powered vehicles

System theory, design concepts, simulation and experimental evaluation

von der Fakultät
Energie-, Verfahrens- und Biotechnik
der Universität Stuttgart
zur Erlangung der Würde eines
Doktors der Ingenieurwissenschaften (Dr.-Ing.)
genehmigte Abhandlung

vorgelegt von
Matthias Rink
geboren in Tübingen

Hauptberichter: Prof. Dr.-Ing. Ulrich Nieken
Mitberichter: Prof. Dr.-Ing. Frank Opferkuch

Datum der mündlichen Prüfung:
20.03.2014

Institut für Chemische Verfahrenstechnik
der Universität Stuttgart

2014

Bibliografische Information der Deutschen Nationalbibliothek

Die Deutsche Nationalbibliothek verzeichnet diese Publikation in der Deutschen Nationalbibliografie; detaillierte bibliografische Daten sind im Internet über http://dnb.d-nb.de abrufbar.

ISBN 978-3-8325-3683-1

Logos Verlag Berlin GmbH
Comeniushof, Gubener Str. 47,
10243 Berlin
Tel.: +49 (0)30 42 85 10 90
Fax: +49 (0)30 42 85 10 92
INTERNET: http://www.logos-verlag.de

Acknowledgments

This thesis summarizes five years of work at the Institute of Chemical Process Engineering. During this period of time, I had the opportunity to work with and learn from numerous collaborators to whom I would like to express my deep gratitude. First, I would like to kindly thank my doctoral supervisor Prof. Ulrich Nieken for giving me both freedom and guiding support for developing and pursuing creative ideas. Special thanks go to Prof. Gerhart Eigenberger for many fruitful discussions and his permanent interest in my work. It was his stalwart optimism and encouragement which spurred me to success! I also owe great thanks to Prof. Frank Opferkuch for his co-advisorship and Prof. Manfred Piesche for chairing the doctoral examination. I really appreciate the special working atmosphere at the institute during all these years. Winni, sharing an office with you has been a particular pleasure. Thank you for the many inspiring discussions about all the world and his wife. Besides talking, concerted sessions of messing around with technical equipment in the laboratory were great fun! Lunch breaks were significantly upgraded by innumerable foosball matches with Carlos, Christian, Franz and Stefan. I also wish to thank my numerous student helpers which all contributed to the success of this work. Generous support, both financial and technical, during the EU-Project InGas is gratefully acknowledged. Many thanks to Dr. Michel Weibel and team (Daimler AG), Dr. Fadil Ayad (Delphi), Janusz Puczok (Katcon) and Dr. Kauko Kallinen (Ecocat) for many inspiring discussions and their great support. I also owe great thanks to Alois Fürhapter (AVL) for pursuing engine testing of the upscaled heat-exchanger prototypes. Last but certainly not least, I would like to thank my loving wife Cornelia for her patience and believe in the final success of this thesis. By cheering me up with her big smile, my little daughter Lilia taught me that there are things in life more important than a doctoral thesis.

Reutlingen, spring 2014

Contents

Symbols and abbreviations 9

Zusammenfassung 15

Abstract 21

1. Introduction 25
1.1. Engine concepts and specific challenges for exhaust purification 26
1.2. Heat-integrated exhaust purification 27
1.3. Thesis objectives and structure 30

2. Simulation models 31
2.1. 1D-multiphase simulation models 31
2.1.1. Heat-exchanger reactor 31
2.1.2. Reference system 38
2.2. Simplified mathematical models for stationary analysis 38
2.2.1. Introduction and motivation 38
2.2.2. Quasihomogeneous models for exemplary design cases 38

3. Stationary simulations 53
3.1. Parameter continuation and stability analysis in DIANA 54
3.2. Analysis of stationary operating behavior 55
3.2.1. Standard ceramic monolith with catalytic coating 55
3.2.2. Fully coated heat-exchanger concept 57
3.2.3. Partially coated heat-exchanger concept 64
3.3. Continuation of design specifications 65
3.3.1. Standard ceramic honeycomb 67
3.3.2. Heat-integrated systems 69
3.4. Conclusions 72

4. Dynamic simulations 73
4.1. Transient behavior of partially coated heat exchanger 74
4.1.1. Heating in countercurrent operating mode 74

4.1.2. Heating with flap/bypass system . . . 76
4.2. Heating strategies for operation under drive cycle conditions . . . 82
4.2.1. Feed conditions for drive cycle simulations . . . 82
4.2.2. Geometric properties of full-scale systems . . . 82
4.2.3. Bypass/flap system, no auxiliary heating . . . 84
4.2.4. Bypass/flap system with electric auxiliary heating . . . 89
4.3. Sequential system . . . 93
4.3.1. Geometric properties of full-scale sequential prototype . . . 94
4.3.2. Comparative study on NEDC performance . . . 96
4.4. Conclusions . . . 99

5. Reactor prototypes and experimental evaluation **103**
5.1. Folded sheet prototype . . . 104
5.1.1. Reactor layout and dimensions . . . 104
5.1.2. Stationary experimental evaluation . . . 106
5.2. Brazed prototype . . . 113
5.2.1. Reactor layout and dimensions . . . 113
5.2.2. Stationary experimental evaluation . . . 115
5.2.3. Comparison of heat exchanger performance . . . 117
5.2.4. Stationary stoichiometric conditions . . . 117
5.3. Sequential system - experimental evaluation . . . 122
5.3.1. Reactor layout and dimensions . . . 123
5.3.2. Transient cold start experiments . . . 127
5.3.3. Stationary experiments . . . 131
5.4. Conclusions . . . 133

6. Directions for future work **135**

Bibliography **141**

A. Experimental Facilities **149**
A.1. Test rigs for experimental evaluation of heat-exchanger prototypes . . 149
A.1.1. Experimental setup for stationary, fuel-lean conditions . . . 149
A.1.2. Experimental setup for stationary, stoichiometric conditions . . 150
A.1.3. Experimental setup for transient, stoichiometric conditions . . . 152
A.2. Exhaust gas generator . . . 153

B. Derivation of quasihomogeneous model equations **157**
B.1. Catalytically coated, countercurrent reactor . . . 157
B.2. Standard ceramic monolith . . . 163

C. Approximation of light-off temperatures **167**
C.1. Ignition time 168
C.2. Estimation of ignition temperatures in case of multiple reactions 169

D. Geometric and thermophysical properties of simulation models **171**
D.1. Geometric properties of different reactor prototypes 171
D.1.1. Folded-sheet prototype 171
D.1.2. Brazed prototype 174
D.1.3. Ceramic monolith 176
D.2. Properties of gas phases and solid materials 177
D.2.1. Gas density 177
D.2.2. Heat capacity and enthalpy 177
D.2.3. Viscosity 178
D.2.4. Heat conductivity 179
D.2.5. Diffusion coefficients 180
D.2.6. Properties of reactor materials 181
D.3. Transport parameters 181
D.3.1. Axial dispersion of heat and mass 181
D.3.2. Coefficients for heat and mass transfer 181

Symbols and abbreviations

Latin symbols

A	m^2	cross-sectional area
a	$\frac{m^2}{m^3}$	volume-specific surface area
a_F	-	amplification factor
c	$\frac{mol}{m^3}$	molar concentration
c_p	$\frac{kJ}{kg \cdot K}$	mass-specific heat capacity
$cpsi$	$\frac{\text{cells}}{\text{in}^2}$	cells per square inch
D	$\frac{m^2}{s}$	coefficient of diffusion / dispersion
d_c	$\frac{\text{cells}}{m^2}$	metric cell density
d_h	m	hydraulic diameter
E_a	$\frac{kJ}{mol}$	activation energy
F	-	bypass flap position
$GHSV$	$\frac{1}{h}$	gas hourly space velocity
h	m	height
h	$\frac{kJ}{mol}$	specific enthalpy
k	$\frac{kmol}{m^2 \cdot s}$	kinetic rate constant
k^0	$\frac{kmol}{m^2 \cdot s}, \frac{1}{s}$	pre-exponential factor
L	m	length
L_p	-	limit point
mil	$\text{in} \cdot 10^{-3}$	1/1000 inch
MW	$\frac{kg}{kmol}$	molecular weight

$\dot{m}$	$\frac{kg}{m^2 \cdot s}$	mass flux density
p	$Pa, bar, mbar$	total or partial pressure
$\mathbf{R}$	$\frac{J}{mol \cdot K}$	universal gas constant
r	$\frac{kmol}{m^2 \cdot s}$	reaction rate
s	m	metal sheet / wall thickness
s_F	-	scaling factor
t	s	time
t_s	s	bypass switch time
T	$°C, K$	temperature
u_g	$\frac{m}{s}$	gas velocity
w	-	weight fraction
w	m	width
y	-	molar fraction
z	m	spatial coordinate

Greek symbols

α	$\frac{W}{m^2 \cdot K}$	heat transfer coefficient
β	$\frac{m}{s}$	mass transfer coefficient
γ	-	dimensionless activation energy
Γ	-	dimensionless length of uncoated part in heat exchanger
Δh_R	$\frac{kJ}{mol}$	reaction enthalpy
ΔT_{ad}	K	adiabatic temperature rise
ϵ	-	volume fraction
ζ	-	dimensionless spatial coordinate
η	-, %	heat exchanger efficiency
η	$\frac{kg}{m \cdot s}$	dynamic viscosity

Θ	-	dimensionless temperature
λ	-	air-fuel ratio ("Lambda")
λ	$\frac{W}{m \cdot K}$	heat conductivity
ν	-	stoichiometric coefficient
ξ	-	drag coefficient
ρ	$\frac{kg}{m^3}, \frac{mol}{m^3}$	density
τ	$s, \frac{1}{s}$	residence time, space velocity
χ	-	conversion
Ψ	-	distribution factor of adiabatic heat release

Lower-case indices

ax	axial
ch	channel
cs	compression jacket
eff	effective
f	channel wall
i	reaction index
ign	ignition
j	component index
k	channel index (1 for inflow, 2 for outflow channels)
sc	side channel
sp	spacer
tot	total
0	initial conditions

Upper-case indices

amb	ambience
c	spacer
c, s	between spacer and wall
ext	external
g	gas
g, c	between gas and spacer
g, s	between gas and wall
in	inflow
N	norm conditions (i.e. $T = 273.15\ K$ and $p = 1.01325\ bar$)
out	outflow
s	wall
s, amb	between wall and ambience
$*$	quasihomogeneous phase

Dimensionless numbers

NTU	number of transfer units
Nu	Nusselt number
Pe	Péclet number
Pr	Prandtl number
Sc	Schmidt number
Sh	Sherwood number

Abbreviations

CNG	Compressed Natural Gas
$CRT^{®}$	Continuously Regenerating Trap
$CSTR$	Continuous Stirred-Tank Reactor

DOC	Diesel Oxidation Catalyst
DPF/DSF	Diesel Particulate/Soot Filter
EGR	Exhaust Gas Recirculation
GWP	Global Warming Potential
HC	Hydrocarbon
NEDC	New European Drive Cycle
NSC	NO_x Storage Catalyst
SCR	Selective Catalytic Reduction
COBYLA	**C**onstrained **O**ptimization **B**y **L**inear **A**pproximation
DIANA	**D**ynamic **s**imulation **an**d **N**onlinear **A**nalysis
PROMOT	**Pro**cess **Mo**deling **T**ool

Zusammenfassung

3-Wege-Katalysatoren werden als effektive Standardlösung für die Entfernung verschiedener Schadstoffe aus dem Abgas Lambda-geregelter Ottomotoren eingesetzt. Neben der Notwendigkeit zum oszillierenden Wechsel der Abgaszusammensetzung zwischen oxidierenden und reduzierenden Bedingungen stellt insbesondere der rasche und emissionsarme Kaltstart eine Herausforderung dar. Während es bei Saugmotoren ausreicht, die Abgastemperatur nach dem Start durch verspätete Zündung anzuheben und so für einen raschen Light-Off der Oxidationsreaktionen am Katalysator zu sorgen, führt die Entwicklung verbrauchsoptimierter Verbrennungsmotoren mit Turbolader und Abgasrückführung zu einem späteren Erreichen der erforderlichen Light-Off Temperaturen. Insbesondere die Turbine verursacht durch ihre thermische Trägheit eine deutliche Verzögerung des Aufheizvorgangs. Kompensiert wird dies in der Regel durch eine möglichst motornahe Anordnung des Katalysators, sowie die Verwendung dünnwandiger Substrate. Eine weitere, interessante Alternative stellen elektrische Heizelemente dar, welche sich im Zuge der bereits begonnenen Hybridisierung von Antriebsträngen in Zukunft leichter umsetzen lassen werden. Bisher wurde ein solches System lediglich in einem relativ kleinen Rahmen im Oberklasse-Segment serienmäßig eingesetzt [46]. Allen aktiven Heizmaßnahmen liegt jedoch das Problem zugrunde, dass im Falle konstant niedriger Abgastemperaturen permanenter Heizbedarf besteht, was den Kraftstoffverbrauch entsprechend erhöht.

Das Institut für Chemische Verfahrenstechnik verfolgt daher seit einigen Jahren das Konzept der wärmeintegrierten Abgassysteme. Dabei wird das zuströmende Motorabgas gemäß Abbildung 1.1 in einem Gegenstromwärmeübertrager vom abströmenden Abgas aufgeheizt, wobei die erforderliche Wärme am Abgaskatalysator freigesetzt oder zusätzlich am heißen Ende eingetragen wird. In einer ersten Arbeit wurde gezeigt, dass sich ein Gegenstromwärmeübertrager mit integriertem Oxidationskatalysator und/oder Partikelfilter hervorragend für die Entfernung brennbarer gasförmiger Bestandteile sowie Ruß eignet [56]. In diesem Fall finden die chemischen Reaktionen auf einem optimalen Temperaturniveau statt ($\sim 700°C$), obwohl die Einströmtemperatur niedrig ($\sim 200°C$) und die treibende Temperaturdifferenz klein ist (~ 100 K). Zudem konnte gezeigt werden, dass mit diesem System die kontinuierliche Regenerierung des Rußfilters mit NO_2 ohne zusätzlichen Kraftstoffmehrverbrauch realisiert werden kann.

In einer weiteren Arbeit wurde der Ansatz des integrierten Gegenstromreaktors weiterverfolgt, wobei insbesondere der Einsatz keramischer Monolithe als Wärmetauscher im Fokus stand [5]. Im Gegensatz zu metallischen Apparaten weisen diese geringere spezifische Massen auf, sodass Wärmeübertrager mit relativ niedrigem Gewicht realisiert werden können. Außerdem lassen sich keramische Substrate generell besser katalytisch beschichten als metallische. Durch partielle Versiegelung der ansonsten porösen Kanalwände des Monolithen lässt sich zudem ein Wall-Flow-Filter (mit porösen Wänden) im Gegenstromwärmeübertrager (mit abgedichteten Wänden) integrieren. Die thermische Beständigkeit der Wandversiegelung erwies sich jedoch als relativ niedrig. Als weiterer Aspekt wurde in dieser Arbeit der gezielte Wärmeeintrag während des Kaltstarts über einen Kraftstoffbrenner untersucht. Das System erlaubt zudem durch partielle, katalytische Oxidation des Kraftstoffes die Bereitstellung eines H_2-/CO-reichen Gasgemisches zur Beschleunigung des Light-Off, sowie zur Regenerierung eines optional vorhandenen NO_x-Speichers. In beiden Arbeiten wurden die Wärmetauschersysteme vor allem hinsichtlich ihres stationären Betriebsverhaltens und für die Reinigung von Dieselabgas untersucht.

Im Rahmen dieser Arbeit steht die wärmeintegrierte Abgasreinigung von Fahrzeugen mit monovalent betriebenen Erdgasmotoren im Fokus. Ein Großteil der Arbeit erfolgte im Rahmen des EU-Forschungsprojektes "InGas - Integrated Gas Powertrain". Die Besonderheit liegt in diesem Fall darin, dass sich Methan aufgrund seiner hohen thermodynamischen Stabilität nur schwer katalytisch umsetzen lässt. Typischerweise werden hierfür Temperaturen zwischen 350°C und 400°C benötigt. Im Gegensatz zu früheren bivalenten Motorkonzepten, bei denen sämtliche Heizmaßnahmen des Katalysators sowie die Zertifizierung im Benzinbetrieb durchgeführt wurden, läuft der Motor im monovalenten Fall permanent im Gasbetrieb. Dies erfordert das möglichst rasche Erreichen des Light-Off von Methan am Katalysator mit anschließender Stabilisierung des Temperaturniveaus. Diese Zielsetzung führte zu einer stärkeren Fokussierung auf den dynamischen Betrieb wärmeintegrierter Abgasreinigungssysteme.

Als Basis dieser Untersuchungen werden zunächst in Kapitel 2 verschiedene mathematische Simulationsmodelle eingeführt, die im Laufe der Arbeit Anwendung finden. Die Untersuchung des dynamischen Betriebsverhaltens wird mit einem 1D-Multiphasenmodell vom Konvektions-/Diffusionstyp durchgeführt. Desweiteren werden verschiedene quasihomogene Modelle vorgestellt, die, obwohl deutlich vereinfacht, die wesentlichen physikalischen Einflussparameter in sehr konzentrierter Form enthalten. Die gute Übereinstimmung mit dem detaillierten Ansatz wird anhand vergleichender Stationärergebnisse verdeutlicht.

Zunächst werden die Vorteile wärmeintegrierter Systeme im Vergleich zu Standardkatalysatoren mit Hilfe umfangreicher Simulationsstudien in Kapitel 3 theoretisch begründet. Als Fallbeispiel dient die Entfernung von CO (niedrige Light-Off-Temperatur) und Methan (hohe Light-Off-Temperatur) aus einem Abgasstrom für drei unterschiedliche Reaktorkonfigurationen. Als Referenzsystem wird im ersten Fall ein Standardkatalysator ohne Wärmeübertrager angenommen. Das resultierende Betriebsfenster für den sicheren Umsatz von Methan ist in diesem Fall auf Einströmtemperaturen oberhalb der Light-Off-Temperatur beschränkt. Anschliessend wird ein Gegenstromwärmetauscher mit durchgehender Katalysatorbeschichtung untersucht. In diesem Fall wird eine deutliche Stabilisierung der Reaktionen, auch bei niedrigen Abgastemperaturen, beobachtet. Zudem treten interessante nichtlineare Effekte auf, die anhand von Simulationsrechnungen mit kontinuierlicher Parameterfortsetzung illustriert und diskutiert werden. Als weitere Verbesserung der Stationärperformance, speziell für den Fall niedriger bis mittlerer Einströmtemperaturen und Raumgeschwindigkeiten, wird im letzten Fall ein Gegenstromwärmetauscher mit partieller Beschichtung untersucht.

Während in den Kapiteln 2 und 3 der Stationärbetrieb im Vordergrund stand, wird in Kapitel 4 nach geeigneten transienten Betriebsstrategien gesucht. Das träge Aufheizverhalten wird zunächst anhand dynamischer Simulationen mit konstanten Zulaufbedingungen demonstriert. Im Anschluss wird eine Strategie präsentiert, bei der das Abgas während der Heizphase über einen Bypass direkt in den katalytisch beschichteten Teil und anschließend durch die Ausströmkanäle des Wärmeübertragers geleitet wird. In diesem Fall erreicht der Katalysator rasch die erforderliche Anspringtemperatur. Nach Beenden des Aufheizvorgangs wird in den Gegenstrombetrieb gewechselt, wobei die Katalysatortemperatur im weiteren Verlauf stabilisiert und weiter erhöht werden kann. Neben den Vorteilen des beschleunigten Kaltstarts wird durch die Bypassklappe zudem der Hochlastbetrieb verbessert. Sobald die Abgastemperatur gewisse Grenzen überschreitet öffnet die Klappe wieder, wodurch die Verstärkungswirkung des Wärmeübertragers deaktiviert und der Gegendruck aufgrund der geraden Durchströmung abgesenkt wird.

Die entwickelte Strategie wird anhand von dynamischen Simulationen mit realistischen Rohemissionsdaten eines Erdgasfahrzeuges weiter verfeinert und der Reinigungsleistung eines Standardsystems ohne Wärmetauscher gegenübergestellt. Im Standardfall sinkt die Katalysatortemperatur während jeder Leerlaufphase des betrachteten Fahrzyklus unter die erforderliche Light-off Temperatur, wodurch in den darauf folgenden Beschleunigungsphasen signifikanter Methanschlupf auftritt. Mit

Wärmeintegration bleibt die Temperatur hingegen wesentlich stabiler, sodass die Reaktion nach erfolgter Zündung zu Beginn des Fahrzyklus nicht wieder verlischt.

Eine weitere Beschleunigung des Zündvorgangs kann durch den Einsatz eines elektrischen Heizelements (E-Kat) erreicht werden. Dies ermöglicht zudem, je nach maximaler Leistung des Heizelements, den Wärmeübertrager weiter entfernt vom Motor zu platzieren und weitestgehend autark zu betreiben.

Obwohl sich das wärmeintegrierte System mit Hilfe geeigneter Betriebsstrategien relativ schnell aufheizen lässt, ist der Temperaturanstieg während der Heizphase im Vergleich zu einem herkömmlichen Katalysator mit keramischem Substrat wesentlich langsamer. Aus diesem Grund wird in einem letzten Schritt ein sequentielles System präsentiert, bei dem ein herkömmlicher keramischer Katalysatorblock mit einem inerten metallischen Wärmeübertrager, sowie mit dem zuvor eingeführten Bypass kombiniert wird. Durch diese Anordnung werden die Nachteile während des Kaltstarts erfolgreich kompensiert, was durch vergleichende Simulationsrechnungen mit dem zuvor gezeigten integrierten System unterstrichen wird. Zudem erlaubt dieser Ansatz die Verwendung herkömmlicher Katalysatorsubstrate. Der Aufbau wird hierdurch deutlich vereinfacht. Spezielle Herausforderungen bei Design und Betrieb des Systems werden durch vergleichende Simulationen diskutiert und es werden entsprechend angepasste Betriebsstrategien vorgestellt.

Nachdem die für die Arbeit wesentlichen Reaktorkonzepte, sowie deren charakteristisches Betriebsverhalten unter stationären und dynamischen Bedingungen anhand von Simulationsergebnissen vorgestellt wurden, erfolgt in Kapitel 5 die Präsentation entsprechender experimenteller Ergebnisse. Zu Beginn wird der Aufbau und die Stationärperformance zweier integrierter Prototypen im Labormaßstab dargestellt und diskutiert. Der erste Prototyp basiert auf dem im Vorfeld dieser Arbeit entwickelten Faltreaktorprinzip, bei dem entsprechend niedrige Materialstärken realisiert werden können. Somit lassen sich verhältnismäßig leichte Wärmeübertrager bauen. Allerdings ist die mechanische Stabilität des resultierenden Kanalpakets eher niedrig und der Aufwand des Fertigungsprozesses hoch. Deshalb wurde im Falle des zweiten Prototypen gemeinsam mit industriellen Partnern ein gelötetes Flachrohr-Design entwickelt, welches sich vergleichsweise einfach fertigen lässt. Beide Prototypen wurden mit dem selben 3-Wege-Katalysator beschichtet und unter mageren Abgasbedingungen getestet. Ferner wurden die gewonnenen Ergebnisse zur Validierung des 1D-Multiphasenmodells verwendet. Hierbei wird deutlich, dass sich die in der heißesten Zone der Wärmeübertrager auftretenden Wärmeverluste deutlich auf die Gesamtperformance auswirken. Dieser Effekt konnte durch eine Begleitbeheizung, beziehungsweise verbesserte Isolierung des heißen Ende des Apparates, kompensiert werden. Insgesamt fallen die Ergebnisse beider Prototypen recht ähnlich aus, wobei die Neuentwicklung deutliche Vorteile beim Gegendruck bietet.

Im letzten Schritt wurde ein Laborprototyp des zuletzt entwickelten sequentiellen Systems aufgebaut und getestet. Ziel dieser Untersuchung war insbesondere, die Kaltstartstrategie mit Heizer und Bypassklappe experimentell zu demonstrieren. Dazu wurde der Reaktor mit einem elektrischen Heizer, entsprechender Spannungsversorgung sowie pneumatisch betätigten Abgasklappen versehen. Mit diesem Aufbau wurden Kaltstartversuche auf einem dynamischen Teststand durchgeführt. Um möglichst realitätsnahe Gaszusammensetzungen zu erhalten wurde ein Rekuperatorbrenner verwendet, der nahezu stöchiometrisches Abgas bei moderaten Temperaturen liefert. Diesem Gasstrom kann CO, sowie ein in seiner Stärke oszillierender Methanstrom zugemischt werden. Auf diese Weise lassen sich reale Abgaszusammensetzungen und optimale Betriebsbedingungen für den verwendeten 3-Wege-Katalysator realisieren.

Die Ergebnisse der Heizversuche, sowie des anschließend untersuchten Stationärbetriebes, bestätigen die im Zuge der zuvor durchgeführten Simulationsrechnungen gewonnenen Ergebnisse.

Zum Abschluss der Arbeit wird ein Ausblick auf mögliche, zukünftige Anwendungsfälle für das sequentielle, wärmeintegrierte Abgasreinigungssystem gegeben. Hierbei geht es zum Einen um mager betriebene Erdgasmotoren, die neben einem Oxidationskatalysator eine weitere Reinigungsstufe für die Entstickung des Abgases benötigen. Zum Anderen wird ein System vorgestellt, das sich für die Behandlung partikelhaltiger Dieselabgase eignet und einige entscheidende Vorteile hinsichtlich Betriebskosten und Reinigungseffizienz bietet. Verglichen mit dem zuvor untersuchten integrierten System [5] ermöglicht die Verwendung standardisierter Katalysatorblöcke zudem einen deutlich vereinfachten Aufbau.

Abstract

Three-Way-Catalysts are a well-established solution for exhaust purification of λ-controlled spark ignition engines. Besides the need for an oscillating switch between fuel-rich and -lean exhaust gas composition, achieving a rapid coldstart with few emissions is specifically challenging. In naturally aspirated engine concepts, delayed ignition usually leads to an immediate rise of exhaust gas temperature and fast light-off, resulting in very low cold start emissions. However, recent improvements of engine technology (e.g. turbo charger and exhaust gas recirculation) lead to generally lower exhaust teperatures and delayed heating during cold start. Especially the turbine's thermal inertia is responsible for this effect which impinges on cold start emissions. The most common countermeasures are a close-coupled position of the catalyst brick and the application of substrates with very low wall thickness. Another interesting option is the use of electric heaters, which could become increasingly facilitated by the hybridization of power trains. So far, serial application of this system was limited to a rather small amount of F-segment vehicles [46]. The general drawback of all heating strategies mentioned above is the fact, that in case of constantly low exhaust temperatures permanent heating support is required. This inevitably leads to increased fuel consumption.

The Institute of Chemical Process Engineering has been pursuing the approach of heat-integrated exhaust purification for several years. According to Figure 1.1, the inflowing cold exhaust is heated in a countercurrent heat exchanger by the hot effluent. The required (small) heat input is either released at the catalyst or optionally applied at the heat exchanger's hot end. In a first work, the potential of a countercurrent heat exchanger with integrated oxidation catalyst and/or soot filter for purification of exhaust flows containing gaseous or solid pollutants was demonstrated [56]. In that case, chemical reactions occur under optimal conditions ($\sim$ 700°C), despite rather low values of feed temperature ($\sim$ 200°C) and adiabatic temperature rise at the catalyst ($\sim$ 100 K). Moreover, the system's potential for the continuous soot filter regeneration with NO_2 was demonstrated. Contrary to state-of-the-art systems without heat recuperation, this does not lead to an increase of fuel consumption.

In a second work, the integrated countercurrent reactor concept was further pursued, focussing especially on the use of ceramic monoliths as heat-exchanger substrates [5]. Due to the low density of those materials, very light heat exchangers can be realized. Moreover, adhesion between washcoat and wall is considerably improved. By partial sealing of the porous channel walls, a countercurrent heat exchanger (sealed walls) with integrated wall-flow filter (porous walls) can be achieved. However, the thermal resistance of the sealing material proved to be rather poor. For rapid cold start, a fuel burner was envisaged. By partial catalytic oxidation of the fuel, H_2-/CO-rich exhaust is supplied which leads to decreased light-off temperatures. Moreover, these species are beneficial during regeneration of an optional NO_x trap. In both works described above the primary goal was to characterize the heat-exchanger systems under stationary conditions.

This work aims at providing heat-integrated exhaust purification concepts for monovalent natural gas powered vehicles. The major part was carried out within the scope of the EU-funded project "InGas - Integrated Gas Powertrain". The key issue in this case is the relatively high light-off temperature ($\sim$ 350°C to 400°C) of residual methane, contained in the exhaust of these vehicles. In contrast to previous, bivalent engine concepts, where critical operating phases were carried out in gasoline mode, monovalent engines have to permanently operate in gas mode. As a result, the catalyst has to quickly reach and maintain temperatures above methane light-off. This work therefore focusses especially on the dynamic operation of heat-integrated exhaust purification systems.

Different simulation models are introduced in Chapter 2 in order to provide a basis for the theoretical analysis presented in the course of this work. Dynamic operation is evaluated with a 1D-multiphase model of convection / diffusion type. Additionally, several quasihomogeneous models, containing the essential physical model parameters in a very condensed form, are presented. The good agreement between these models and the detailed approach is illustrated by comparative stationary results.

In Chapter 3, the advantages of heat-integrated systems over standard catalysts are illustrated based on a comprehensive simulation study. As exemplary case, the removal of CO (low light-off temperature) and methane (high light-off temperature) is evaluated assuming three different reactor configurations. Initially, a standard catalyst without heat exchanger is applied. The resulting operating range for complete conversion of methane is restricted to regions of sufficiently high exhaust temperature (above methane light-off). Subsequently, a countercurrent heat exchanger with catalytic coating over the complete channel length is evaluated. In this case the reactions are stabilized, with the operating range extending to exhaust temperatures considerably below methane light-off. Moreover, interesting nonlinear effects can be observed, which are illustrated and discussed by parameter continuation techniques. For even improved stationary performance, especially in case of low to medium feed temperatures and space velocities, a partially coated system is proposed.

While Chapters 2 and 3 predominantly deal with stationary operation, in Chapter 4 appropriated operating strategies for dynamic conditions are introduced. The slow transient behavior of countercurrent heat exchangers is illustrated by dynamic simulations with constant feed conditions. Then, a strategy is presented in which during heating the exhaust enters the heat exchanger directly at the coated part and exits though the outflow channels. In this case, the required light-off temperature is quickly reached. After sufficient heating the system operates in countercurrent mode, leading to a significant stabilization of catalyst temperature. Besides improved cold start performance, re-opening the bypass during high engine load helps to avoid excessive temperatures and backpressure.

The strategy is further refined based on dynamic simulations with realistic raw emission data of a natural gas powered vehicle. In case of a standard system without heat exchanger, the catalyst temperature drops below the required light-off range during each idling phase. This leads to significant slip of methane during subsequent acceleration of the vehicle. By heat integration however, the catalyst temperature can be stabilized after having achieved ignition of methane during the initial phase of the drive cycle.

Even shorter heating periods can be realized by adding an electric heater to the system. Depending on its maximum heating power output, this setup allows for autonomous operation of the heat-integrated system, independent of engine-out conditions.

Although the heating behavior can be considerably improved by the strategies described above, the temporal gradient of catalyst temperature is considerably lower compared to a standard system with ceramic catalyst substrate and equal cell density. As a result, a sequential layout, based on a combination of ceramic monolith catalyst, metallic heat exchanger and the bypass system, is introduced. The improved performance of this system especially during cold start is underlined by additional drive cycle simulations. Since in this case standard ceramic monoliths can be applied as catalyst substrates, the system layout can be significantly simplified. Specific challenges regarding optimized design and operating strategies of these systems are discussed based on simulation results.

After having presented the main reactor concepts with characteristic stationary and dynamic operating behavior based on simulation results, in Chapter 5 experimental results are shown. Initially, stationary results obtained with two integrated prototypes at laboratory scale are discussed. The first prototype is based on the previously developed folded-sheet design which allows for very low wall thickness. This leads to low weight of the resulting heat exchangers. However, the folded-sheet structure is relatively flexible, leading to poor mechanical stability of the applied washcoat. Moreover, the manufacturing process is rather complex. As a result, for the second prototype a vacuum-brazed design approach, based on a stack of flat tubes, was developed together with industrial partners. With the brazing process finally optimized, this manufacturing technique is well suited for automated mass customization. Both heat exchangers were coated with the same Three-Way-Catalyst and tested under fuel-lean conditions. The resulting stationary axial profiles of temperature and concentration were used for fitting of the 1D-multiphase model. Heat losses to ambience evidently impinge on the thermal performance of the systems. As a result, the hot end of the heat exchangers was fitted with improved insulation and compensation heating. On the whole, both prototype generations behave quite similarly under stationary conditions with significant benefits of the brazed system regarding backpressure.

In the last step, a laboratory-scale prototype of the sequential system was built and tested. Special focus was put on demonstrating the cold start strategy with bypass flap and heater. As proposed in the simulation part of this work (Chapter 4), the system was equipped with a small electric heater and appropriate power supply. For the bypass system, pneumatically actuated flaps were fitted. This setup was then tested on a dynamic test rig. For realistic exhaust composition, a recuperative gas burner was applied. This system generates nearly stoichiometric exhaust at moderate temperatures ($< 300°C$). In order to mimic the exhaust composition of a stoichiometric engine, CO and oscillating fluxes of methane were added. These conditions are required for optimal catalyst performance.

The experimental results obtained with the sequential system under both transient and stationary conditions confirm the principal findings of the previously performed simulation study.

In the last Chapter of this work, an outlook is given on future applications of the sequential system. First, a layout for lean gas engines is presented which contains an additional catalytic De-NO_x stage. Subsequently, a setup for particle-loaded diesel exhaust is presented which is expected to exhibit some significant benefits regarding operating costs and purification performance. Compared to the previously developed, integrated system [5], the application of standard ceramic catalyst substrates leads to a significantly simplified system setup.

Chapter 1.

Introduction

In the medium term, the energy supply of automobile vehicles is likely to change due to the gradual depletion of actual crude oil reservoirs and increasingly difficult exploitation of new ones. As one of the most promising alternatives readily available, natural gas has slowly come to the fore during the past several years. While being already extensively used in Latin America and Asia, natural gas powered vehicles have played a rather negligible role in European countries [21]. Over the recent years, the amount of natural gas available on the world market has increased considerably which leads to decreasing prices especially in the United States [72]. There, a growing share of gas is actually exported due to booming domestic production. The low price compared to diesel or gasoline makes the fuel interesting especially for fleet vehicles, such as delivery trucks, urban buses or taxis, since they do not depend on a widespread gas infrastructure. By a stepwise change from liquid to gaseous fuels, natural gas could also serve as transition fuel, allowing for admixtures of fossil gas, biogas and/or hydrogen [30]. In the long term, natural gas could also be used to power fuel cells leading to improved energetic efficiency compared to internal combustion engines.

However, a considerable decrease of CO_2 emissions could already be achieved with state-of-the-art engine technology due to the low carbon content of the methane molecule. While there is no doubt regarding the benefits in terms of lower CO_2 emissions, the CNG technology is actually still rather pricey compared to top-grade gasoline or diesel engines which are also capable of reaching well-to-wheel emissions between 120 and 140 g CO_2 per kilometer [70]. The disadvantage arises from the high cost of components required for CNG operation (i.e. high-pressure tanks, injectors, special catalytic converters), as well as of the additional gasoline equipment which serves as fallback option. As a result, the vehicles oftentimes do not amortize for private owners with average annual mileages of $\approx$16000 km. Yet, with improving gas infrastructure the gasoline components are not required any longer. In combination with increased lot sizes this allows for cheaper production. With respect to more stringent emission legislation targeting well below 120 g/km CO_2, the actually available CNG technology is favorable even under consideration of current oil prices [70]. Moreover, the European Union is likely to penalize violation of the emission limits which would inevitably lead to a further rise in price of conventionally fueled cars and give a competitive edge to CNG vehicles. Certainly, improvements of CO_2 emissions can also be achieved by hybridization of powertrains. Yet, this technology is no option for small to medium sized vehicles due to the relatively high price of batteries and control equipment required.

Despite the numerous advantages of CNG as a fuel, the high thermodynamic stability of the methane molecule implicates several drawbacks. First of all, residual methane is difficult to remove from the exhaust of gas powered vehicles due to the high temperature required for its catalytic conversion. Second, if emitted the molecule remains in the atmosphere for a relatively long time span. Due to its high capacity to absorb heat radiation from the earth's surface, it is beleived to exhibit high global warming potential (GWP relative to $CO_2 \approx 25$ to 27 [7, 22] on a time horizon of 100 years). Therefore, safe removal of methane from industrial and automotive exhaust is a challenge to be faced alongside with increasing use of CNG as fuel.

1.1. Engine concepts and specific challenges for exhaust purification

The engines of natural gas powered passenger cars currently available on the market usually operate under stoichiometric conditions and are optimized on CNG as fuel. In order to enhance vehicle range and provide a fallback solution in case of insufficient natural gas infrastructure, small gasoline tanks are fitted, allowing the engine to automatically change the fuel supply if necessary. In these so-called MONOVALENTPLUS vehicles the driver has no longer any influence on the operating mode of the engine. Even cold start and catalyst heating has to be performed in CNG mode. Compared to former naturally aspirated CNG engines with port fuel injection, modern concepts are turbo charged which significantly enhances volumetric efficiency of the engine [73]. This leads to a certain drop of exhaust temperature since the turbine acts as heat sink. The exhaust is even colder if the engines are equipped with (cooled) exhaust gas recirculation (EGR) [20]. By dilution of the combustion mixture with inert exhaust, the temperature in the cylinders decreases. This leads to lower heat loss and higher engine efficiency. Additionally, the engine's resistance against knocking improves, allowing for higher compression ratios and increased engine efficiency. In fact, the fuel itself is very stable regarding self-ignition with octane numbers ranging between 105-122 (real mixtures) and 128 (pure methane) [73]. However, the drawback of increased EGR is that, especially during low-load operation, the exhaust temperature after turbine drops well below the values required for complete removal of residual methane [20].

Effects similar to those described above for stoichiometric operation with EGR can be obtained by running the engine with lean air-fuel mixtures [20]. In fact, under certain operating conditions engine efficiency is even higher than in the stoichiometric case. However, this leads to increased NO_x emissions. While they can be easily removed with a three-way-catalyst and reducing gas atmosphere, this is not possible in the lean case. Instead, a combination of NO_x trap and oxidation catalyst is required to simultaneously remove all pollutants. An interesting strategy to reduce NO_x emissions of both stoichiometric and lean-burn engines during combustion in the cylinders is to use mixtures of hydrogen and natural gas as fuel [17]. Due to the improved combustion efficiency this concept also leads to a significantly lower amount of residual methane in the exhaust. In the medium to long term with increasing availability of hydrogen, this option could become realistic. At the moment however, the well-to-wheel efficiency of this so-called HCNG is clearly inferior to diesel fuel.

As conclusion, modern CNG engines, specifically optimized on the new fuel, reach or even exceed the performance figures of comparable gasoline or diesel engines. However, the low exhaust temperatures arising from application of top-line engine technology are oftentimes very challenging in terms of safe removal of hydrocarbon emissions. This leads to relatively high catalyst volume and/or increased precious metal content in order to undercut the increasingly stringent emission limits.

1.2. Heat-integrated exhaust purification

While intensive research on catalytic oxidation of methane has been performed for quite a long time without yielding significant improvements with respect to decreased light-off temperatures, in this work a different conceptual approach is presented. The basic idea originates from the concept of autothermal operation of fixed-bed reactors where heat exchangers have been commonly applied in order to preheat the feed with the hot effluent of the reactor. Even the small amount of heat released by oxidation of a highly diluted pollutant is sufficient to maintain the required temperature level for catalytic oxidation if an efficient recuperative or regenerative heat exchanger is used [18, 49, 52]. Without the heat exchanger significantly higher energy inputs would be required in order to stabilize the catalyst temperature at the same level. In standard autothermal reactor concepts a countercurrent heat exchanger is coupled with an adiabatic fixed-bed reactor. An alternative comprises a countercurrent reactor where the hot part of the countercurrent section is coated with the oxidation catalyst. In Figure 1.1, top, a schematic of the last-mentioned approach

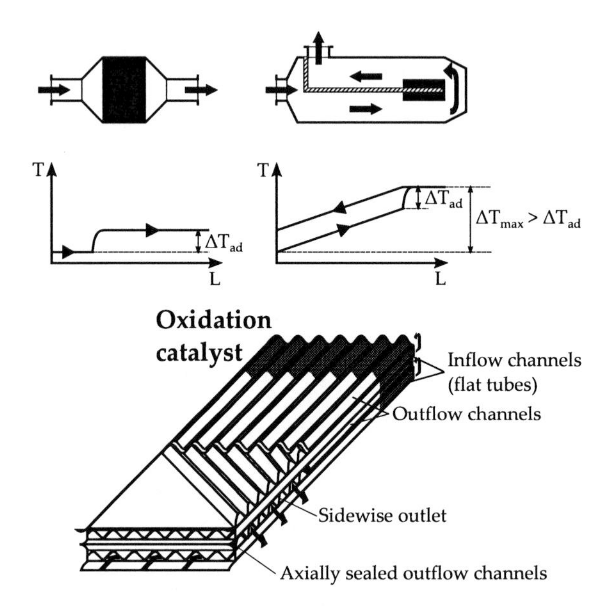

Figure 1.1.: Comparison of ideal axial temperature profiles in standard system (top left) and heat-integrated case (top right) where a certain zone of a countercurrent heat exchanger is catalyst coated. An exemplary setup of a parallel-plate, countercurrent heat exchanger layout is depicted below.

for exhaust purification is shown and compared to an adiabatic fixed-bed reactor. Under ideal conditions the increase of gas temperature between inlet and outlet of the catalytic converter corresponds to the adiabatic temperature rise ΔT_{ad}, which depends on the inflow concentration of the reacting gas species (Figure 1.1, top left). In case of the heat-integrated system (Figure 1.1, top right), the exhaust passes first through the inflow channels of a countercurrent heat exchanger where it is heated by the hot outflowing gas passing in opposite direction through the adjacent channels. When reaching the catalyst section, the same amount of heat as in the standard case is released, the gas flows back and leaves the heat exchanger sidewise. Due to the recovery of reaction heat, the catalyst temperature is significantly higher than in the standard case. Previously published results show that in a laboratory-scale setup the maximum temperature rise can reach values which are about eight times higher than the adiabatic one [5]. There, and in a second preceding work [56], research focussed on aftertreatment of exhaust gas from diesel engines. As reactor concepts, ceramic or metallic countercurrent heat exchangers [24, 25] were developed and tested. In Figure 1.1, bottom, the metallic type layout is exemplarily shown since the ceramic option was not further pursued in this work. In the present case, the heat exchanger package is composed of a stack of flat tubes made of thin metal sheets. Corrugated spacer structures are placed within for enhanced mechanical stability and improved heat transfer. The exhaust gas enters these channels at the left-hand side in axial direction. Between the flat tubes, corrugated spacers are placed which serve as outflow channels with the gas passing through in opposite direction. In order to achieve countercurrent flow, the axial faces of these channels are sealed at the inflow end. A short slice with tilted corrugation is applied to achieve a sidewise outlet. The remaining lateral channel faces have to be gas tight in order to avoid internal leakage between inflow and outflow channels. For the metallic heat exchanger different manufacturing concepts can be applied. In this work, prototypes based on the previously developed folded-sheet design [24, 25] and according to a new, brazed layout are built up and tested. After completion of the heat exchanger core, the catalyst is applied by dip-coating the heat exchanger's right end over a certain axial length. Finally, the system is insulated with ceramic fibre materials and canned.

For natural gas vehicles also a regenerative approach based on a fxed-bed reactor with periodic flow reversal is proposed in [67]. This system is able to achieve full conversion of methane even with the feed temperature set to ambient conditions. Yet, in neither of the previous works appropriate strategies for fast system startup or high-load operation were proposed. Moreover, especially the flow-reversal approach is rather difficult to implement with regard to rapidly changing feed conditions. As a result, this work aims at providing a fundamental background of both the stationary and the dynamic operational characteristics of heat-integrated systems. In addition, system layouts and operating strategies optimized for application in natural gas vehicles are presented.

1.3. Thesis objectives and structure

In Chapter 2, a 1D-multiphase model of the heat-integrated exhaust converter is introduced which is applied in Chapter 4 for transient simulations and development of dynamic operating strategies. In addition, dimensionless quasihomogeneous simulation models are introduced and characteristic numbers are derived which reflect the influence of different design and operating parameters on system performance during stationary operation.

The main goal of Chapter 3 is to elucidate the system-theoretical background of different heat-integrated design approaches in more detail, based on nonlinear analysis of the respective quasihomogeneous models. This study aims at demonstrating the beneficial effect of the countercurrent heat exchanger on system performance in terms of reliable pollutant conversion under severe conditions (i.e. exhaust temperatures well below pollutants' light-off temperatures). Moreover, the impact of different design and operating parameters on system performance is investigated systematically by direct continuation of certain design specifications.

After having discussed design layouts suitable for stationary operation, dynamic operating strategies based on the detailed multiphase model are presented in Chapter 4. The significance of a bypass of the heat exchange section for rapid catalyst heating is illustrated first by transient simulations assuming constant feed conditions and subsequently with dynamic raw emission data of a CNG powered vehicle. These results serve as basis for the development of the final design approach which comprises a sequential alignment of a ceramic honeycomb catalyst brick and an inert metallic heat exchanger. Comparative drive cycle simulations of this new concept and of the previous fully integrated system are used to demonstrate the advantageous effect of reduced thermal mass in the catalytic part.

In Chapter 5, the different reactor prototypes developed and built in the course of this work are presented together with specific experimental results. First, the stationary performance of two different metallic heat exchangers with integrated catalytic coating is discussed. Finally, transient and stationary tests performed with a lab-scale sequential system are shown.

Chapter 6 gives a brief outlook on modified layouts of the sequential system for lean-burn CNG engines and for particle-loaded diesel exhaust.

Chapter 2.

Simulation models

This chapter contains the mathematical models used for simulation of the different heat-integrated concepts and the reference system. Initially, the detailed approach applied for transient simulations (see Chapter 4) and validation of stationary experimental results (see Section 5.1.2 and Section 5.2.2) is shown. Related model parameters are listed in Appendix D. Subsequently, simplified quasihomogeneous models are introduced which were used for the more system-theoretical studies of the heat-integrated systems and the deduction of design principles (see Chapter 3).

2.1. 1D-multiphase simulation models

The general equations of the 1D-multiphase heat exchanger model originate from [29] and were implemented in the PROMOT/DIANA simulation tool [39, 40]. Additional relations required for modeling of specific features and phenomena during transient operation were included and are described in Chapter 4.

2.1.1. Heat-exchanger reactor

In the following, the model equations used for simulation of the countercurrent heat-exchanger reactor are shown. They are based on the balance element depicted in Figure 2.1. The heat-exchanger reactor is a parallel plate reactor, composed of thin metal sheets, where corrugated spacers separate adjacent sheets and increase the heat exchange surface between gas and wall. In the catalyst section the complete package (i.e. spacers and wall) is catalytically coated. The model comprises two gas phases (inflow and outflow channel) with opposite flow directions. Additionally, the spacers at both sides and the wall are treated separately, resulting in five phases for the complete balance element. The mass flux density refers to the total cross-sectional area A_{tot}. In Appendix D the required geometric and thermophysical properties are listed.

The balance equations contain the following terms:

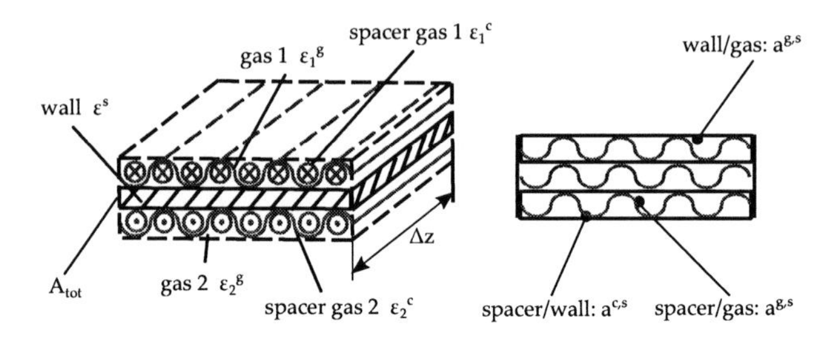

Figure 2.1.: Balance element for simulation model (left) with different types of heat and mass transfer (right).

- Accumulation terms for the properties stored in a volume element $A_{tot}\Delta z$.
- Convective transport terms for properties entering the volume element through the front surface $\epsilon A_{tot}|_z$ and leaving it through the back surface $\epsilon A_{tot}|_{z+\Delta z}$.
- Transport terms for exchange with neighboring phases assuming linear driving force.
- Source terms for reactive generation and consumption in the volume element (only within catalytically active zones).

The following simplifications apply:

- Changes of properties within one phase occur only in axial flow direction. This assumption is justified by the clear domination of axial convection over radial diffusion in the gas phases and the low thickness of solid phases.
- In each channel three phases interact: the flowing gas, the spacer structures and the separating walls to the adjacent channels. Every phase is considered by its respective volume fraction ϵ_k and by its specific exchange area $a_{v,k}$.
- In the gas phases mass transfer in flow direction takes place by convection and dispersion. In the solid phases axial heat conduction is the only transport mechanism.
- For the gas phases laminar plug flow is assumed and axial pressure drop is considered.
- The solid properties are constant in time and space.

- The gas phases are ideal.
- The same mass transfer coefficient applies for all gas species.
- Heat and mass exchange between the phases occurs according to linear transfer equations. The analogy between heat and mass transfer is postulated ($Pr \approx Sc$). The corrugated spacer structures act as fins and therefore enhance heat transfer between gas and solid. This effect is accounted for by a linear driving force term between spacers and wall with constant heat transfer coefficient and specific contact surface.
- Heat loss to the environment is accounted for by a linear driving force term in the wall phase energy balance with constant heat transfer coefficient and the outer surface area of the complete heat exchanger package.
- Heat loss effects caused by radiation are neglected.

Additionally, the following assumptions are made for the integration of catalytic combustion kinetics into the heat exchanger model:

- In the material balance coated areas of wall and spacer structures form one homogeneous phase.
- The material balance of the solid phase is quasi-stationary.
- Due to the small exchange fluxes between gas and solid, enthalpy changes are neglected in the energy balance equations.
- Catalytic combustion of methane and CO is modeled with basic power law kinetics since the reliable prediction of the respective light-off temperature is regarded sufficient to assess the thermal performance of the heat-exchanger concepts.

2.1.1.1. Energy balance of gas phases $\left[kJ/m_{tot}^3/s\right]$

The gas phase energy balance accounts for heat exchange with the spacer structures (index c) and the wall (index s) in the respective channel k. The sign in front of the convection term depends on the flow direction (i.e. - for inflow channels ($k = 1$), + for outflow channels ($k = 2$)).

$$\begin{aligned} \epsilon_k^g \cdot \rho_k^g \cdot c_{p,k}^g \cdot \frac{\partial T_k^g}{\partial t} &= \mp \dot{m}_k \cdot c_{p,k}^g \cdot \frac{\partial T_k^g}{\partial z} + \epsilon_k^g \cdot \lambda_{ax}^g \cdot \frac{\partial^2 T_k^g}{\partial z^2} + \\ &+ a_{v,k}^{g,c} \cdot \alpha_k^{g,c} \cdot \left(T_k^c - T_k^g\right) + a_{v,k}^{g,s} \cdot \alpha_k^{g,s} \cdot \left(T^s - T_k^g\right) \end{aligned} \tag{2.1}$$

Since equation 2.1 is of second order in space, two boundary conditions are required. They are formulated as a pair of *Danckwerts* boundary conditions [50].

$$\dot{m}_k \cdot c_{p,k}^g \cdot \left(T_k^{in} - T_k^g \left(z = 0/L\right)\right) = -\epsilon_k^g \cdot \lambda_{ax}^g \cdot \left.\frac{\partial T_k^g}{\partial z}\right|_{z=0/L} \tag{2.2}$$

$$\left.\frac{\partial T_k^g}{\partial z}\right|_{z=L/0} = 0 \tag{2.3}$$

Depending on the simulation task (i.e. transient cold start simulations or stationary calculations) either ambient temperature or an axial temperature profile is applied as initial condition for the gas temperatures.

$$T_k^g \left(t = 0, z\right) = T_k^{g,0} \left(z\right) \tag{2.4}$$

2.1.1.2. Energy balance of corrugated spacers $\left[kJ/m_{tot}^3/s\right]$

In the spacer phases, energy transport occurs by axial heat conduction, heat exchange with both the surrounding gas phase and the wall and heat generation by catalytic combustion at the coated surface. Since the exact specific surface of the washcoat is unknown, the reaction rate refers to the free geometric surface of the spacer structures.

$$\begin{aligned}\epsilon_k^c \cdot \rho_k^c \cdot c_{p,k}^c \cdot \frac{\partial T_k^c}{\partial t} &= \epsilon_k^c \cdot \lambda_k^c \cdot \frac{\partial^2 T_k^c}{\partial z^2} + a_{v,k}^{g,c} \cdot \left(T_k^g - T_k^c\right) + \\ &+ a_{v,k}^{c,s} \cdot \alpha_k^{c,s} \cdot \left(T^s - T_k^c\right) + a_{v,k}^{g,c} \cdot \sum_i \left(-\Delta h_R\right)_i \cdot r_{i,k}\end{aligned} \tag{2.5}$$

Both ends of the reactor are assumed to be adiabatic which leads to *Neumann* boundary conditions.

$$\left.\frac{\partial T_k^c}{\partial z}\right|_{z=0,L} = 0 \tag{2.6}$$

As initial condition, the same temperature as for the initial condition of the gas phases is assumed.

$$T_k^c \left(t = 0, z\right) = T_k^g \left(t = 0, z\right) \tag{2.7}$$

2.1.1.3. Energy balance of the wall $\left[kJ/m_{tot}^3/s\right]$

The main transport mechanisms in this balance equation are axial heat conduction, heat exchange with the adjacent gas phases and the spacer structures which are assumed to touch the wall at either side. Additionally, heat is generated at the catalyst coating at both sides of the wall. Heat losses to the environment refer to the outer geometric surface of the heat exchanger package whose temperature is assumed to be equal to the wall temperature.

$$\begin{aligned} \epsilon^s \cdot \rho^s \cdot c_p^s \cdot \frac{\partial T^s}{\partial t} &= \epsilon^s \cdot \lambda^s \cdot \frac{\partial^2 T^s}{\partial z^2} + \sum_k \left(a_{v,k}^{g,s} \cdot \alpha_k^{g,s} \cdot (T_k^g - T^s) + \right. \\ &+ \left. a_{v,k}^{c,s} \cdot \alpha_k^{c,s} \cdot (T_k^c - T^s) \right) + a_v^{s,amb} \cdot \alpha^{s,amb} \cdot \left(T^{amb} - T^s \right) + \\ &+ a_{v,k}^{g,s} \cdot \sum_i (-\Delta h_R)_i \cdot r_{i,k} \end{aligned} \tag{2.8}$$

As for the spacers, adiabatic ends are assumed and the initial temperature is set equal to the gas phase temperature:

$$\left. \frac{\partial T_k^s}{\partial z} \right|_{z=0,L} = 0 \tag{2.9}$$

$$T_k^s (t=0, z) = T_k^g (t=0, z) \tag{2.10}$$

2.1.1.4. Material balance of gas phase components $\left[kg/m_{tot}^3/s\right]$

As for the gas phases' energy balance equations, the main transport mechanisms are assumed to be convection by the gas flow, mass transfer to the spacer and wall surface and axial dispersion. As simplification, the mass transfer coefficient is equal for all components.

$$\begin{aligned} \epsilon_k^g \cdot \rho_k^g \cdot \frac{\partial w_{j,k}^g}{\partial t} &= \mp \dot{m}_k \cdot \frac{\partial w_{j,k}^g}{\partial z} + \epsilon_k^g \cdot \rho_k^g \cdot D_{ax} \cdot \frac{\partial^2 w_{j,k}^g}{\partial z^2} + \\ &+ \left(a_{v,k}^{g,c} + a_{v,k}^{g,s} \right) \cdot \beta \cdot \rho_k^g \cdot \left(w_{j,k}^s - w_{j,k}^g \right) \end{aligned} \tag{2.11}$$

As for the gas phases' heat balance equations, *Danckwerts* boundary conditions are assumed [50]:

$$\dot{m}_k \cdot \left(w_{j,k}^{in} - w_{j,k}^g (z = 0/L) \right) = -\epsilon_k^g \cdot \rho_k^g \cdot D_{ax}^g \cdot \left. \frac{\partial w_{j,k}^g}{\partial z} \right|_{z=0/L} \tag{2.12}$$

$$\left. \frac{\partial w_{j,k}^g}{\partial z} \right|_{z=L/0} = 0 \tag{2.13}$$

The system is initialized with axially constant gas composition:

$$w^{g}_{j,k}\left(t=0,z\right)=w^{g,0}_{j,k} \tag{2.14}$$

2.1.1.5. Material balance of components at the catalyst surface $\left[kg/m^{3}_{tot}/s\right]$

For the quasi-stationary material balance of the catalyst phases, reaction at the surface sites and mass exchange with the surrounding gas phases is assumed:

$$0=MW_{j}\cdot\left(a^{g,c}_{v,k}+a^{g,s}_{v,k}\right)\cdot\sum_{i}\nu_{i,j}\cdot r_{i,k}-\left(a^{g,c}_{v,k}+a^{g,s}_{v,k}\right)\cdot\rho^{g}_{k}\cdot\beta\cdot\left(w^{s}_{j,k}-w^{g}_{j,k}\right) \tag{2.15}$$

For enhanced numerical stability during initialization of the system, the catalyst phase gas composition is set equal to the one in the gas phases:

$$w^{s}_{j,k}\left(t=0,z\right)=w^{g}_{j,k}\left(t=0,z\right)=w^{g,0}_{j,k} \tag{2.16}$$

2.1.1.6. Axial pressure drop

In order to simulate the backpressure caused by laminar flow through the heat exchanger's channels, the following relation is used:

$$\frac{dp_{k}}{dz}=\xi_{k}\frac{\eta_{k}u_{g,k}}{d_{h}^{2}}, \tag{2.17}$$

with the drag coefficients ξ_k, the gas phase viscosities η_k, the gas velocities $u_{g,k}$ and the hydraulic diameter of the channels d_h. Ideal coupling is assumed between inflow (i.e. k = 1) and outflow (i.e. k = 2) channels and the outlet is set to ambient pressure:

$$p_{1}|_{z=L}=p_{2}|_{z=L} \tag{2.18}$$

$$p_{2}|_{z=0}=p_{amb}. \tag{2.19}$$

Since deflection of the flow at the transition between inflow and outflow channels and at the sidewise outlet probably causes additional backpressure, the two phases are treated with separate drag coefficients which include the above mentioned phenomena in a rather lumped way. As described in Section 5.2.2, this approach is able to correctly predict the pressure drop between heat exchanger inlet and outlet during laboratory-scale testing.

2.1.1.7. Kinetic model

The main purpose of the kinetic model applied in this study is to correctly describe the ignition behavior of the two gas phase components CH_4 and CO considered, without excessive computational effort and numerical problems. Therefore, a simple approach based on power-law rate equations was adopted from [29] with the required rate constants fitted accordingly. The resulting equations for $j = CO, CH_4$ read:

$$r_j = k_j(T) \cdot p_j^s, \tag{2.20}$$

$$k_j(T) = k_j^0 \cdot exp\left(\frac{E_{a,j}}{\mathbf{R}T}\right), \tag{2.21}$$

with the partial pressure of the respective component in the solid phase (p_j^s) and the rate constant $k_j(T)$ whose temperature dependence is expressed by an Arrhenius equation. The reaction rates refer to the free geometric surface of the solid (wall and spacers) and therefore have units $\left[kmol_j/m^2/s\right]$. In Table 2.1, the initially applied constants are listed.

Table 2.1.: Original model parameters for catalytic oxidation of CH_4 and CO.

$k_{CO}^0 \left[\frac{kmol_{CO}}{m^2 sbar}\right]$	$E_{a,CO} \left[\frac{kJ}{mol}\right]$	$k_{CH_4}^0 \left[\frac{kmol_{CH_4}}{m^2 sbar}\right]$	$E_{a,CH_4} \left[\frac{kJ}{mol}\right]$
100	55	20	74

Chapter 3 demonstrates that the ignition temperatures obtained with this model agree quite well with values published in literature. Moreover, the model is able to predict the methane conversion during laboratory-scale experiments under fuel-lean conditions with decent accuracy (see Chapter 5). Yet, this approach is certainly not sufficient to capture the complete range of complex phenomena occurring in a natural-gas vehicle's three-way catalyst. In this context, more complex kinetic models, such as a recently published global approach comprising five separate reactions [71], are more appropriate. For example, the high water content in the exhaust of natural gas vehicles leads to a significant decrease of methane conversion activity [12, 26]. Moreover, operation under fuel-lean conditions causes a slow decay of catalyst activity [8, 9, 13]. Very similar effects also emerged in the course of laboratory-scale experiments with the heat-exchanger prototypes of this work (see Chapter 5).

2.1.2. Reference system

For the reference system (i.e. a standard ceramic honeycomb), the same modeling approach as previously described for the countercurrent heat-exchanger reactor was applied. Contrary to the heat exchanger model, only 2 phases (i.e. gas and wall) are considered with the respective mass and energy balance equations. Moreover, the gas flow passes the converter in one direction only. In Section D, the applied model parameters are shown.

2.2. Simplified mathematical models for stationary analysis

2.2.1. Introduction and motivation

The removal of low pollutant concentrations from relatively cold exhaust gas streams has been a challenging task over the past decades. As a result, several designs of autothermal reactors emerged, allowing to increase the temperature level at which the catalytic conversion takes place. One possibility is to ignite the conversion by heating the catalyst bed of a fixed-bed reactor above the pollutant's ignition temperature. If now the flow direction of the cold exhaust through the reactor is repeatedly reversed, the conversion stabilizes [16]. During each phase, the reaction front slowly shifts due to the cooling of the catalyst bed. This so-called reverse-flow reactor has been extensively studied, as reported by [37] and references therein. In Figure 2.2 the basic principle of such a reactor setup is shown.

Over the years, the main challenge regarding mathematical modeling of these systems has been the relatively high computational effort required to attain the stationary profiles of temperature and concentration. Therefore, simplified models were derived for slow and infinitely fast switching periods [44, 45]. In the latter case, the solution obtained with periodic flow-reversal approaches that of a countercurrent reactor, as demonstrated in [18, 53]. This analogy is graphically illustrated by Figure 2.3.

2.2.2. Quasihomogeneous models for exemplary design cases

In the following, quasihomogeneous model equations for the different design cases evaluated in this work are presented. Model development occurred along the lines of similar procedures described in [53] and [23], aiming at concentrating the physical properties in a variety of dimensionless characteristic numbers. In Appendix B, a detailed derivation of the reduced model equations is shown for the most general cases

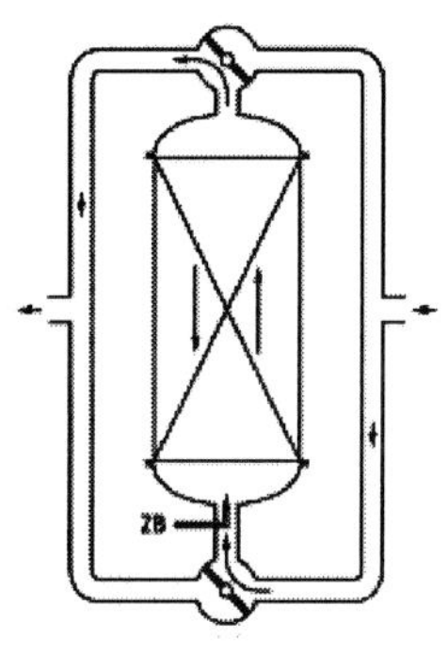

Figure 2.2.: Schematic of adiabatic fixed-bed reactor with valves for periodic flow reversal [37].

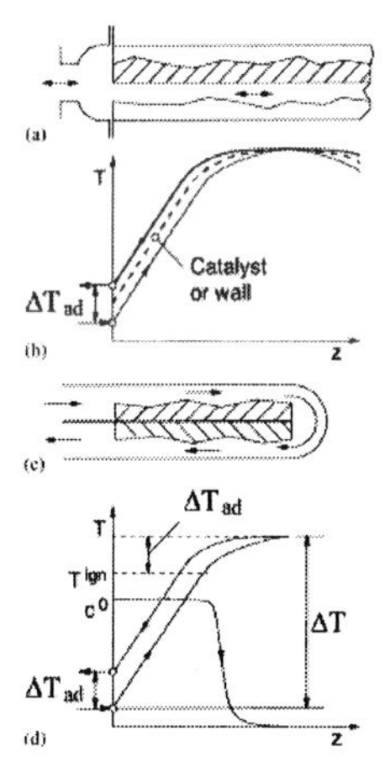

Figure 2.3.: Analogy between periodic flow reversal (a) with short switching periods and countercurrent operation (b) with equal geometric and catalytic properties. The respective axial profiles are shown in (b) and (d) [37].

of a fully coated heat-exchanger reactor and a standard ceramic monolith which is regarded as reference case. This chapter contains the final model equations of four different reactor configurations. Moreover, stationary results obtained with both the quasihomogeneous and the detailed models are shown and compared for each of the observed cases.

2.2.2.1. Fully coated metallic heat exchanger

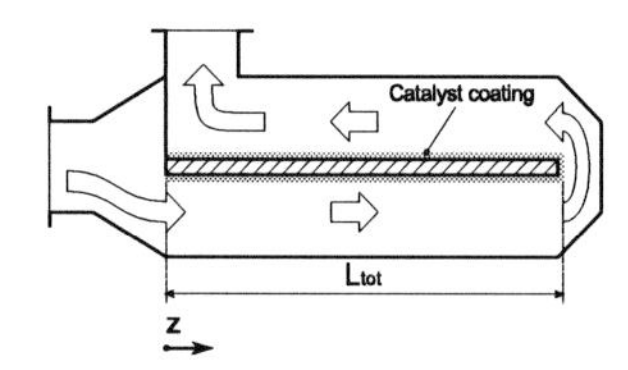

Figure 2.4.: Schematic of fully coated concept.

The first and most general case of a heat-integrated reactor concept considered in this study is a heat exchanger with catalytic coating over the complete channel length. Figure 2.4 shows a schematic of this setup. The cold exhaust gas enters the reactor at the left end and passes through the inflow channels. On its way to the right end of the reactor, the gas flow heats up due to the hot outflowing gas which passes through the adjacent channels in opposite direction. Since catalytic coating is applied over the complete length of the device, chemical reactions can take place at any axial position. As described in Appendix B, a quasihomogeneous stationary energy balance can be derived for this approach, following the principles derived in [53] and [23]:

$$0 = \lambda_{eff} \cdot \frac{d^2 T^*}{dz^2} + \frac{a_v}{2} \cdot \sum_{i=1}^{I} (-\Delta h_{R,i}) \cdot (r_{i,1} + r_{i,2}), \tag{2.22}$$

with the effective heat conductivity λ_{eff} [37, 49, 53, 74]:

$$\lambda_{eff} = (1 - \epsilon) \cdot \lambda^s + \left(\dot{m} c_{p,g}\right)^2 \cdot \frac{4}{\alpha a_v}. \tag{2.23}$$

This parameter reflects the interaction of the different heat transport mechanisms. While the first term of Equation 2.23 represents axial heat conduction in the solid material, the second term reflects heat exchange between the gas flow and the wall.

In order to attain an even more compact form, equation 2.23 can be rearranged, yielding:

$$\lambda_{eff} = \dot{m}c_{p,g}L_{tot} \cdot \left(\underbrace{\frac{(1-\epsilon)\cdot\lambda^s}{\dot{m}\cdot c_{p,g}\cdot L_{tot}}}_{\widehat{=}Pe^{-1}} + \underbrace{\frac{4\cdot\dot{m}\cdot c_{p,g}}{\alpha\cdot a_v\cdot L_{tot}}}_{\widehat{=}NTU^{-1}} \right). \tag{2.24}$$

Obviously, the two heat transport effects can be described by the characteristic numbers Pe and NTU. After further simplifications and rearrangements (see Appendix B for details), the final energy balance reads:

$$0 = \left(\frac{1}{Pe} + \frac{1}{NTU}\right)\cdot\frac{d^2T^*}{d\zeta^2} + \sum_{i=1}^{I}\Delta T_{ad,i}\cdot\left(\frac{d\chi_{i,1}}{d\zeta} - \frac{d\chi_{i,2}}{d\zeta}\right), \tag{2.25}$$

with the dimensionless length $\zeta = z/L_{tot}$, the adiabatic temperature rise for the i-th reaction $\Delta T_{ad,i}$ and the conversion $\chi_{i,k=1,2}$ of $i = I$ linearly independent components in the inflow ($k = 1$) and outflow ($k = 2$) channels. The only reactions considered in the present case are the catalytic conversion of CO and CH_4. The total adiabatic temperature rise can be therefore calculated as (see Section B.1):

$$\Delta T_{ad,tot} = \sum_{i=1}^{I}\Delta T_{ad,i} = \underbrace{\frac{w_{CH_4,in}\cdot(-\Delta h_{R,CH_4})}{MW_{CH_4}\cdot c_{p,g}}}_{\Delta T_{ad,CH_4}} + \underbrace{\frac{w_{CO,in}\cdot(-\Delta h_{R,CO})}{MW_{CO}\cdot c_{p,g}}}_{\Delta T_{ad,CO}}. \tag{2.26}$$

If the total adiabatic temperature rise in the exhaust and the relation between CO and CH_4 concentration are known, the final energy balance reads:

$$\begin{aligned} 0 &= \left(\frac{1}{Pe} + \frac{1}{NTU}\right)\cdot\frac{d^2T^*}{d\zeta^2} + \Delta T_{ad,tot}\cdot\left[\Psi\left(\frac{d\chi_{CO,1}}{d\zeta} - \frac{d\chi_{CO,2}}{d\zeta}\right) + \right. \\ &+ \left.(1-\Psi)\left(\frac{d\chi_{CH_4,1}}{d\zeta} - \frac{d\chi_{CH_4,2}}{d\zeta}\right)\right], \end{aligned} \tag{2.27}$$

with the respective heat release assigned to either component by the parameter $\Psi := \Delta T_{ad,CO}/\Delta T_{ad,tot}$. This equation is of second order in space and requires two boundary conditions. In [53] and [37] it is shown, that in an ideal countercurrent reactor the temperature difference between inflow and outflow is equal to the adiabatic temperature rise in the reactor. If in addition the quasihomogeneous temperature is assumed to be the mean value between inflow and outflow channel gas temperature, the following relation for the left-hand side boundary can be applied in direct analogy to [53]:

$$T^*|_{\zeta=0} = T_{in} + \frac{\Delta T_{ad,tot}}{2} \cdot \left(\Psi \cdot \chi_{CO,out} + (1-\Psi) \cdot \chi_{CH_4,out}\right), \tag{2.28}$$

with the total conversion values at outflow position $\chi_{CO,out}$ and $\chi_{CH_4,out}$. Since the right end of the reactor is assumed to be insulated, the respective heat flux is zero (see also [37] for comparison) and a *Neumann* condition can be applied:

$$\left.\frac{dT^*}{d\zeta}\right|_{\zeta=1} = 0. \tag{2.29}$$

Together with Equation 2.27, a set of simplified stationary material balance equations is solved. The mass transfer term in the gas phase is eliminated by insertion of the wall material balances (see Appendix B for details). As for the gas phase energy balances, axial dispersion is neglected. The equations for inflow and outflow channel therefore read:

$$0 = \tau \cdot \rho^{N,in} \cdot \epsilon \cdot w_j^{in} \cdot \frac{d\chi_{j,1}}{d\zeta} + MW_j \cdot \frac{a_v}{2} \cdot \sum_{i=1}^{I} \nu_{ij} r_{i,1}, \tag{2.30}$$

$$0 = -\tau \cdot \rho^{N,in} \cdot \epsilon \cdot w_j^{in} \cdot \frac{d\chi_{j,2}}{d\zeta} + MW_j \cdot \frac{a_v}{2} \cdot \sum_{i=1}^{I} \nu_{ij} r_{i,2}, \tag{2.31}$$

with the space velocity τ and gas density $\rho^{N,in}$ at norm conditions and inflow gas composition. As reaction rates, the power law relations introduced in Section 2.1.1.7 are used. Since Equation 2.30 and Equation 2.31 are of first order in space, one respective boundary condition is required. The gas enters the reactor at the left-hand side and therefore a *Dirichlet* boundary condition is applied for the inflow channel side. Inflow and outflow channels are assumed to be ideally coupled at their right end. These assumptions lead to the following two relations:

$$\chi_{j,1}|_{\zeta=0} = 0, \tag{2.32}$$

$$\chi_{j,2}|_{\zeta=1} = \chi_{j,1}|_{\zeta=1}. \tag{2.33}$$

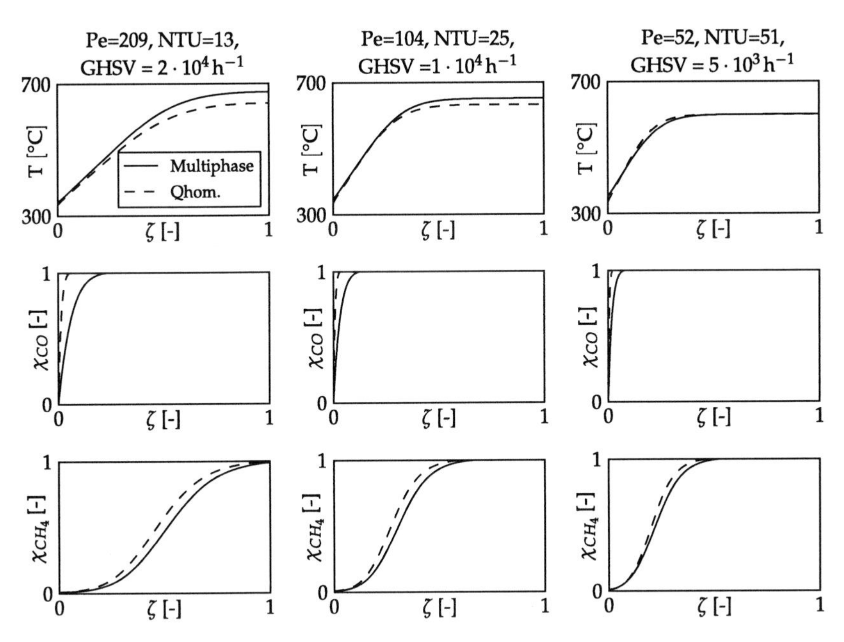

Figure 2.5.: Axial profiles of temperature (top row), CO conversion (centre row) and CH_4 conversion (bottom row) for fully coated concept. Pe, NTU and $GHSV$ are indicated for each column, $\Delta T_{ad} = 60\,K$, $\Psi = 0.2$ and $T_{in} = 300\,°C$ are constant for all three cases.

In order to validate the simplified model, stationary axial profiles of temperature and conversion are shown in Figure 2.5 for three different space velocities and constant feed conditions. For a reasonably fine axial resolution the simplified model agrees well with the detailed approach. The small visible differences arise from the different boundary assumptions and the simplifications made in the quasihomogeneous approach. Due to the significantly different activation energy of the two reactions, a pronounced separation of reaction zones can be observed. While CO reacts spontaneously close to the reactor inlet, CH_4 needs higher reaction temperatures and the reaction front shifts further downstream in the inflow channels with rising $GHSV$. As a result, the maximum temperature remains relatively constant within an optimal range despite the significant variation of $GHSV$. The same effect can be observed if different inflow temperatures are assumed and $GHSV$ is kept constant (Figure 2.6). This self-adaptation effect, which is a unique feature of catalytically coated countercurrent reactors, was studied extensively in the past [6, 44, 49]. In this work, the special case of two simultaneously occurring reactions will be analyzed in more detail (see Section 3.2.2).

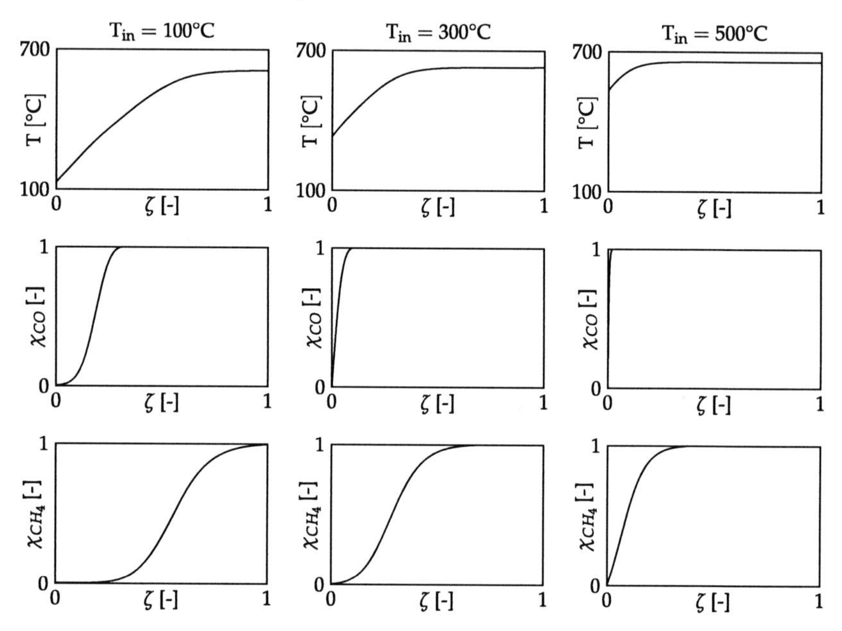

Figure 2.6.: Axial profiles of temperature (top row), CO conversion (center row) and CH_4 conversion (bottom row) for fully coated concept (only solution of quasihomogeneous model shown). The respective value of T_{in} is indicated for each column, $\Delta T_{ad} = 60\,K$, $\Psi = 0.2$ and $GHSV = 10000\,h^{-1}$ are constant for all three cases.

2.2.2.2. Partially coated integrated concept

The consequence of the self-adaptation behavior is, that during stationary operation most of the catalyst is not used. In fact, for reaching full conversion, a very narrow coated zone would be sufficient. Upstream of this zone, the temperature profile is linear, indicating pure heat exchange. The location of the reaction front is determined by the ignition temperature of the respective component. This characteristic behavior is explored graphically in [37].

In this section, a system with limited coating length is introduced and a modified version of the simulation model is presented. In Figure 2.7, a schematic of this setup is shown. While passing through the inert channels of the reactor, the gas is heated

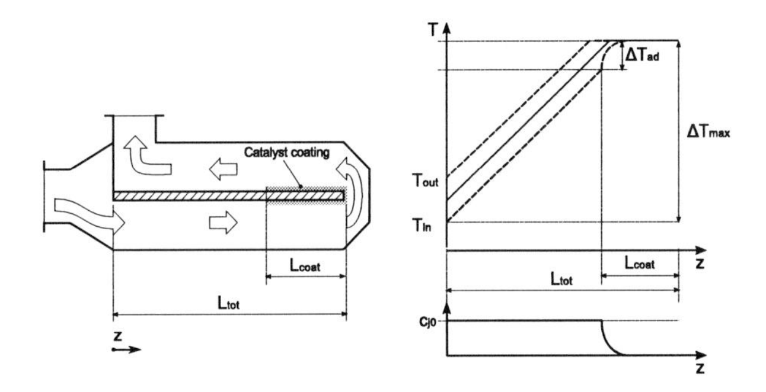

Figure 2.7.: Schematic of partially coated concept approach (left); ideal axial profiles of temperature and concentration (right).

above the ignition temperature of the respective pollutant species. When reaching the catalytic part, the heat of reaction corresponding to the adiabatic temperature rise is released and the profile flattens over the catalytic part as already observed with the fully coated concept. Since the reaction zone is "trapped" further to the right, the catalyst temperature level can be shifted towards significantly higher values. This effect is demonstrated in Section 3.2.3. In order to quantify the heat exchanger's efficiency, the ratio between the maximum temperature rise and the adiabatic one, the so-called "amplification factor" a_F [5, 66], was introduced. It is closely linked to the usually applied heat exchanger efficiency η of the uncoated channels which is defined as the ratio of the actual heat flux exchanged over the channel length to the maximal possible one. If mass flux and heat capacity are assumed to be equal at either channel side, the relation can be simplified using the characteristic temperature differences specified in Figure 2.7, right:

$$\eta = \frac{\Delta T_{max} - \Delta T_{ad}}{\Delta T_{max}} = 1 - \frac{1}{a_F}. \tag{2.34}$$

Hence, optimal efficiency from a theoretical point of view ($\eta = 1$) leads to an infinitely high amplification factor. The values obtained in experimental tests range between 4.5 and 8 [5, 54, 66]. However, the laboratory-scale systems mentioned in these publications suffer from significant heat losses through the external insulation. This is especially severe in case of CNG exhaust purification since the catalyst temperature has to be substantially higher than in diesel systems in order to attain sufficient CH_4 conversion levels.

In the following, the main dependencies of the amplification factor on the derived dimensionless parameters shall be demonstrated. To this end, the dimensionless energy balance (Equation 2.27) is integrated once. Total conversion over the reactor length is assumed in order to obtain the slope of the temperature profile at $\zeta = 0$:

$$\left.\frac{dT^*}{d\zeta}\right|_{\zeta=0} = \frac{\Delta T_{ad,tot}}{\left(\frac{1}{Pe} + \frac{1}{NTU}\right)}. \tag{2.35}$$

The maximum temperature is reached at the beginning of the catalytic part. In order to calculate the dimensionless length of the uncoated part, a new parameter with appropriate limits has to be defined:

$$\Gamma := \frac{L_{uncoat}}{L_{tot}}, \left(\frac{1}{Pe} + \frac{1}{NTU}\right) \leq \Gamma < 1. \tag{2.36}$$

The excluded set

$$\Gamma < \left(\frac{1}{Pe} + \frac{1}{NTU}\right) = \frac{\Delta T_{ad,tot}}{\frac{dT^*}{d\zeta}}$$

refers to conditions where the adiabatic temperature rise is not yet fully released.

Now, the maximum temperature rise can be calculated as:

$$\Delta T_{max} = \left.\frac{dT^*}{d\zeta}\right|_{\zeta=0} \cdot \Gamma = \frac{\Gamma \Delta T_{ad,tot}}{\left(\frac{1}{Pe} + \frac{1}{NTU}\right)}. \tag{2.37}$$

By definition of Γ, ΔT_{max} is always greater than the adiabatic temperature rise $\Delta T_{ad,tot}$. Equation 2.37 implies a relation for the amplification factor:

$$a_F := \frac{\Delta T_{max}}{\Delta T_{ad,tot}} = \frac{\Gamma}{\left(\frac{1}{Pe} + \frac{1}{NTU}\right)}. \tag{2.38}$$

With this result, energy balance 2.27 can be reformulated for the partially coated case as funtion of a_F and Γ:

$$\begin{aligned} 0 &= \frac{\Gamma}{a_F} \cdot \frac{d^2T^*}{d\zeta^2} + \Delta T_{ad,tot} \cdot \left[\Psi \left(\frac{d\chi_{CO,1}}{d\zeta} - \frac{d\chi_{CO,2}}{d\zeta} \right) + \right. \\ &+ \left. (1 - \Psi) \left(\frac{d\chi_{CH_4,1}}{d\zeta} - \frac{d\chi_{CH_4,2}}{d\zeta} \right) \right]. \end{aligned} \tag{2.39}$$

As mentioned above, for values of Γ below the limit defined in Equation 2.36, the general form of the energy balance (Eq. 2.27) has to be applied since the maximum temperature rise can not be calculated with the simple approximation of Equation 2.37.

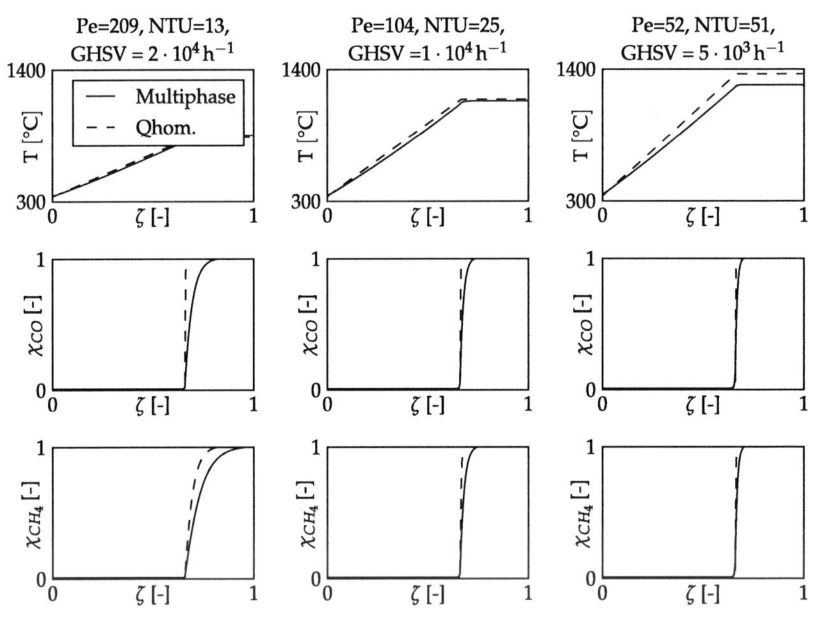

Figure 2.8.: Axial profiles of temperature (top row), CO conversion (center row) and CH_4 conversion (bottom row) for partially coated concept. *Pe*, *NTU* and *GHSV* are indicated for each column, $\Delta T_{ad} = 60\,K$, $\Psi = 0.2$, $\Gamma = 0.67$ and $T_{in} = 300\,°C$ are constant for all three cases.

The material balance equations, as well as all boundary conditions, can be adopted from the previously described case with catalytic coating from end to end. The only difference is that the source terms of both reactions are set to zero in the uncoated part of the heat exchanger. As for the fully coated case (see Figure 2.5), a comparison between the simplified and the detailed approach was performed. The resulting axial profiles are depicted in Figure 2.8. Due to the partially inert heat exchanger, the resulting maximum temperature is much higher than in the fully coated case. In fact, under ideal conditions, the maximum temperature depends linearly on the coated length [37]. As a result of the significantly increased catalyst temperature, both reactions occur at the very beginning of the coated zone. Hence, a separation, as observed in the case of end-to-end coating (Fig.2.5) with self-adaptation of the maximum temperature, is not visible.

2.2.2.3. Sequential system

The final heat-integrated concept developed in this work is based on a sequential alignment of an inert metallic heat exchanger and a standard ceramic monolith. It corresponds to the original autothermal concept where a countercurrent heat exchanger is attached to an adiabatic fixed-bed reactor. A schematic picture of this approach is depicted in Figure 2.9.

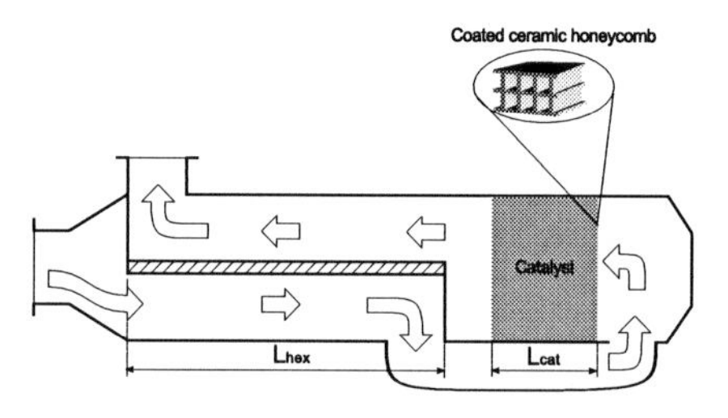

Figure 2.9.: Schematic of the sequential system, comprising a metallic countercurrent heat exchanger and a standard ceramic monolith catalyst.

The quasihomogeneous model of this approach is basically a combination of the previously shown heat exchanger models (Section 2.2.2.1 and Section 2.2.2.2) with a standard ceramic monolith (Section 2.2.2.4). Since the heat exchanger is assumed to be inert, its axial temperature profile is linear under stationary conditions. As a result, Equation 2.35, which describes the slope of the axial temperature profile, can be rearranged in order to obtain an energy balance for the inert heat exchanger:

$$\left(\frac{1}{Pe_{hex}} + \frac{1}{NTU_{hex}}\right) \cdot \frac{dT^*}{d\zeta} = \Delta T_{ad,tot} \cdot \left(\Psi \cdot \chi_{CO,cat,out} + (1-\Psi) \cdot \chi_{CH_4,cat,out}\right), \quad (2.40)$$

with Pe_{hex} and NTU_{hex} according to Equation B.10 and Equation B.11. Since this is a first order problem in space, at the left-hand side a boundary condition has to be formulated. It can be directly adopted from Equation 2.28 with the conversion obtained at the outflow of the catalyst brick. The material and energy balances of the catalyst are identical to the ones derived for the standalone standard system which are presented in the following section. Between both modules, ideal coupling is assumed for both heat and mass fluxes. With this model the effect of asymmetric design configurations (e.g. different values of specific surface a_v in heat exchanger and catalyst) can be simulated. The results of this study are shown in Section 3.3.2. In case of equal geometric parameter choices for catalyst and heat exchanger, the results approach those obtained with the partially coated integrated system. A validation with the detailed model is therefore omitted.

2.2.2.4. Standard ceramic honeycomb

In order to demonstrate the benefits of a heat-integrated exhaust purification system, the previously described cases are compared with the catalytically coated honeycomb structure. This system is regarded as reference case.

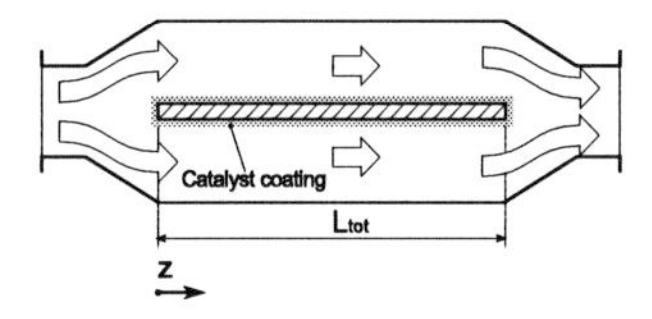

Figure 2.10.: Schematic of standard system.

As can be seen in Figure 2.10, the exhaust gas passes through the catalytically coated channels of the converter in one direction. Consequently, the maximum temperature rise corresponds to the adiabatic one according to the feed gas composition. A detailed derivation of the quasihomogeneous model equations can be found in Section B.2 while here only the final balance equations are given.

First, the energy balance is formulated as

$$
\begin{aligned}
0 &= \left(\frac{1}{Pe_{cat}} + \frac{1}{NTU_{cat}}\right) \cdot \frac{d^2T^*}{d\zeta^2} - \frac{dT^*}{d\zeta} + \\
&+ \Delta T_{ad,tot} \cdot \left[\Psi \frac{d\chi_{CO}}{d\zeta} + (1-\Psi)\frac{d\chi_{CH_4}}{d\zeta}\right],
\end{aligned}
\tag{2.41}
$$

with Pe_{cat} and NTU_{cat} according to Equation B.38 and Equation B.39. Since energy transport to and from the system boundaries is assumed to solely occur by convection, a pair of *Danckwerts* boundary conditions is assigned (see Section B.2 for a detailed derivation):

$$
T^*|_{\zeta=0} = T_{in} + \left(\frac{1}{Pe_{cat}} + \frac{1}{NTU_{cat}}\right) \cdot \left.\frac{dT^*}{d\zeta}\right|_{\zeta=0}, \quad \text{and} \quad \left.\frac{dT^*}{d\zeta}\right|_{\zeta=1} = 0. \tag{2.42}
$$

The material balance in the catalyst can be directly derived from Equation 2.30 and reads:

$$0 = -\tau \cdot \rho^{N,in} \cdot \epsilon \cdot w_j^{in} \cdot \frac{d\chi_j}{d\zeta} + MW_j \cdot a_v \cdot \sum_{i=1}^{I} \nu_{ij} r_i, \tag{2.43}$$

with the inflow boundary condition

$$\chi_j\big|_{\zeta=0} = 0. \tag{2.44}$$

As for the previous cases, the quasihomogeneous model was compared with the detailed solution and the resulting plots are shown in Figure 2.11. These calculations were performed assuming a ceramic monolith with 400 cpsi / 6 mil wall thickness and a total volume of 1.7 l. Together with different material parameters (i.e. solid heat conductivity, heat capacity and density), this leads to different values of Pe and NTU compared to those obtained with the heat-integrated systems at equal $GHSV$.

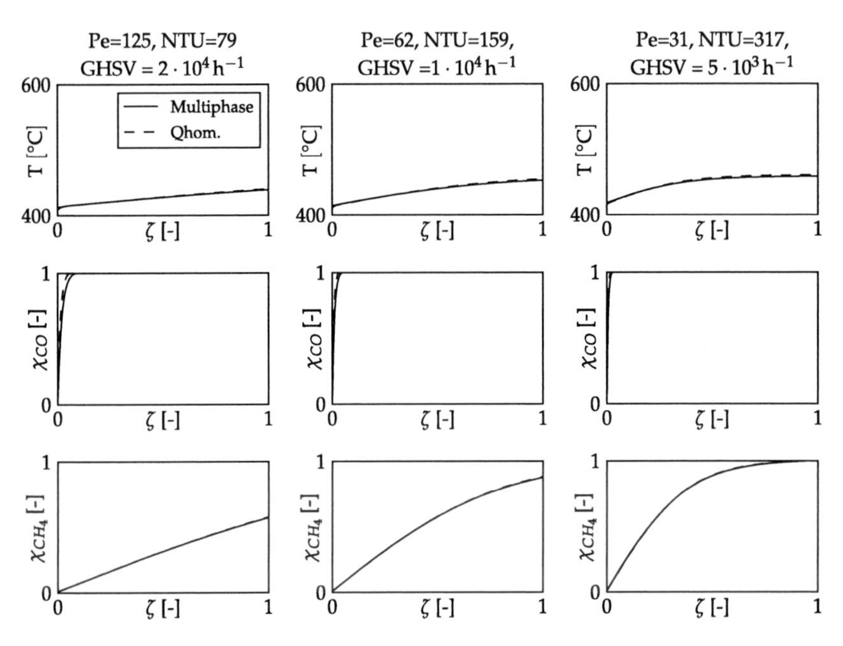

Figure 2.11.: Axial profiles of temperature (top row), CO conversion (center row) and CH_4 conversion (bottom row) for the standard ceramic monolith. *Pe*, *NTU* and *GHSV* are indicated for each column, $\Delta T_{ad} = 60\,K$, $\Psi = 0.2$ and $T_{in} = 400°C$ are constant for all three cases.

Chapter 3.

Stationary simulations

From the derivation of the quasihomogeneous model of the catalytically coated heat-intergrated concept, as described in the preceding chapter, it becomes quite obvious that the stationary system behavior is determined by a limited set of design parameters showing up in the different balance equations. By combining these parameters appropriately, a few dimensionless numbers result which reflect the system's characteristics in an even further condensed and universal form. Therefore, these quasihomogeneous models are highly beneficial to quickly assess the system performance for different design configurations. As output, maps can be generated, displaying the required layout parameters for a given set of design specifications over the complete operating range.

In order to efficiently generate these maps, the simulation software DIANA [39, 40] allows direct parameter continuation of stationary solutions as a function of additional parameters. These techniques were originally developed and applied for stability analysis of nonlinear chemical engineering reactor models. Moreover, direct continuation of higher- dimensional, singular points as a function of model parameters is possible [39, 80, 81, 82]. In this work, up to two parameters are continued simultaneously in order to directly calculate stability limits in the operating parameter space of the observed systems.

After a short introduction on the basic principles of parameter continuation with DIANA, the different simplified model cases introduced in Chapter 2 are studied regarding their operational behavior in the relevant operating range. In order to demonstrate the special benefits of a heat-integrated system with regard to conversion performance under conditions of low exhaust temperature, this evaluation is coupled with a stability analysis of the models. Subsequently, direct continuation of design specifications is used to generate layout maps for the different model cases.

3.1. Parameter continuation and stability analysis in DIANA

With the simulation package DIANA [39, 40], steady-state solutions of a partial differential equation system can be used for parameter continuation tasks. Hence, for a vanishing time derivative, the nonlinear algebraic system to be solved reads:

$$f(x, \lambda) = 0, \tag{3.1}$$

with the state vector $x \in \mathbb{R}^n$ and one free parameter of arbitrary choice $\lambda \in \nu$. Assuming that f is a smooth function with non-singular Jacobian matrix f_x, the continuation could be performed directly by using a Newton-like iteration scheme. However, singular points in parameter space might be crossed that result in a loss of rank in the Jacobian matrix. Therefore, in DIANA a function $f(\zeta)$ with one-to-one correspondence to the original solution at the initial point is solved after appropriate parameterization along the curve-length parameter ζ using predictor-corrector methods. The solution $y(\zeta)$ must therefore satisfy the equation:

$$f(y(\zeta)) = 0, \quad \text{with} \quad y(0) = y_0. \tag{3.2}$$

If the initial point $\{x_0, y_0\}$ is regular, the Jacobian matrix of the parameterized solution curve of Equation 3.2 has maximal rank for all values of ζ along the continuation path. Simultaneously, the eigenvalues of the linearized original system are calculated at each solution point of the continuation. It is well known from control theory, that a system loses stability if one eigenvalue crosses the imaginary axis, which is equivalent to a sign change of its real part. For the case of a completely vanishing eigenvalue, a so-called *critical point* is reached. These points usually indicate a sudden change in system behavior (i.e. ignition/extinction of an exothermic reaction) or a multiplicity of possible solutions. Hence, from a system-theoretical point of view, exact knowledge of these points is of imminent interest since usually the complete variety of possible solutions can be found in their proximity. This methodology is available in DIANA for singularities up to codimension 3 and is demonstrated in [39] for investigating the complex behavior of a continuous stirred-tank reactor (CSTR). An analysis of singularities up to codimension 1 occurring in the solution of the heat-integrated concepts developed in this work are shown in the following section.

3.2. Analysis of stationary operating behavior

3.2.1. Standard ceramic monolith with catalytic coating

As reference system, the model case comprising a standard ceramic honeycomb with catalytic coating was assumed (see Section 2.2.2.4). As geometric layout, a commonly used combination of 400 cpsi cell density and 6 mil ($\sim$ 0.15 mm) wall thickness was adopted. The material properties assumed for the ceramic substrate are documented in Section D.2.6.

The above described parameter continuation technique was used to directly calculate the stationary operating behavior regarding maximum temperature and conversion in the complete window of possible inflow temperatures and constant $GHSV$. In Figure 3.1, the results are shown for a continuation of the stationary solution from fully extinguished to fully ignited state. The obtained plots of conversion over inflow temperature correspond to the result of a classic adiabatic light-off experiment. Compared to methane, ignition of CO occurs at relatively low temperatures

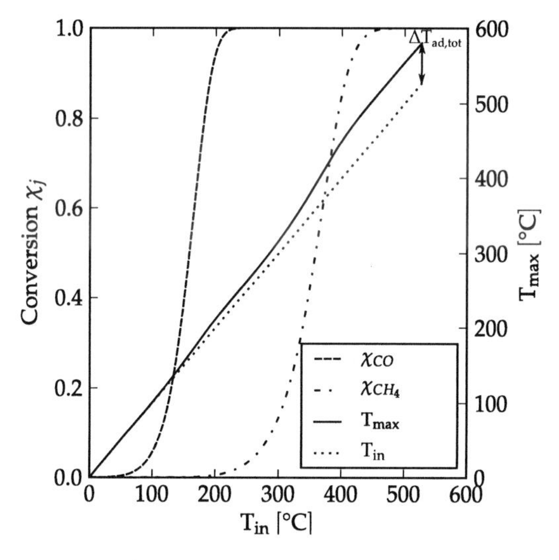

Figure 3.1.: Conversion and maximum temperature over inflow temperature at constant space velocity ($GHSV = 10000\,h^{-1}$) for a standard monolith.

($T_{50,CO} \approx 150 - 200°C$, [3, 36, 55]). Since in this case the adiabatic temperature rise caused by the first reaction is low (12 K), methane conversion is hardly affected. The obtained temperature rise is visible in the small deviation of T_{max} from T_{in}. As soon as methane conversion starts ($T_{50,CH_4} \approx 350 - 400°C$, [34, 36, 48, 55])), the temperature rise is more pronounced until reaching $\sim$ 60 K, corresponding to the value of $\Delta T_{ad,tot}$ when both CO and methane are fully converted. In order to quantitatively estimate the ignition behavior of the catalytic conversion of CO and methane, two relations were formulated in this work describing the so-called "ignition temperature". The equations were directly derived from the gas phase energy balance of an adiabatic fixed-bed reactor (see Appendix C) with the power law kinetics from Section 2.1.1.7 for chemical reactions. In Figure 3.2, the ignition temperatures calculated with Equation C.13 are shown together with the conversion results of Figure 3.1 which were obtained with the quasihomogeneous model. Obviously, the calculated values are very close to the respective T_{50} temperatures and are therefore suitable to forecast the ignition behavior of an exothermic reaction based on a set of appropriate kinetic parameters.

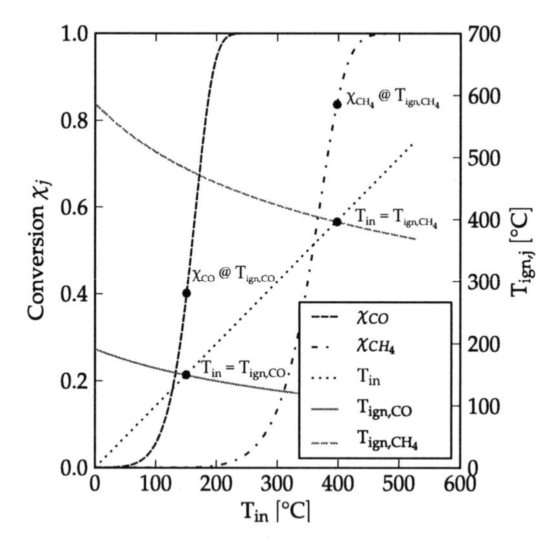

Figure 3.2.: Conversion and adiabatic ignition temperatures at constant space velocity ($GHSV = 10000\, h^{-1}$) for a standard monolith.

For a final evaluation of the standard system's performance in the complete range of possible operating conditions, a map can be generated with the described continuation technique. First, curves of maximum temperature and methane conversion for different constant values of $GHSV$ are calculated using T_{in} as continuation parameter. In the next step, $GHSV$ is continued at different constant values of T_{in}. The resulting three-dimensional surfaces of maximum temperature and methane conversion are depicted in Figure 3.3. Full conversion is obtained at low space velocities and high inflow temperatures. This underlines the classic dilemma of methane abatement from CNG engines: low engine load usually leads to moderate exhaust temperature and mass flow. As a result, catalysts with very high activity and/or specific surface are required in order to achieve sufficiently high conversion levels in this part of the engine operating map. In fact, under currently valid drive cycle conditions, the engine operates in this low and medium load region for most of the time which requires considerable fuel penalties in order to increase and maintain the catalyst temperature.

3.2.2. Fully coated heat-exchanger concept

Before comparing the different heat-integrated design approaches, the fully coated concept is studied in more detail. Although being rather academic due to the high amounts of catalytic material required for complete coating of the relatively big reactor unit, this layout exhibits very interesting operational characteristics from a system-theoretical point of view. Originally, this reactor concept was derived as limiting case of a reverse-flow reactor with infinitely fast switching between the two flow directions [51, 53] due to the tremendously reduced computational cost and the very good agreement of the results with those obtained with the original flow-reversal approach [6, 44]. If a mixture of multiple pollutants with significantly different ignition characteristics is present in the gas flow, multiple steady states develop as studied experimentally in [52]. In the present case, a similar behavior can be expected due to the markedly different light-off temperatures of CO and methane (see Figure 3.2). In Section 2.2.2.1 stationary results are shown which illustrate the self-adaptation effect of the fully coated, countercurrent reactor depending on variations of T_{in} and $GHSV$. In either case, the axial position of reaction zones adapts according to the respective operating conditions. In case of different $GHSV$ and constant T_{in} (Figure 2.5), the axial gradient of the temperature profile decreases with increasing space velocity (see also Equation 2.35). Hence, the temperature level required for ignition of CH_4 is reached and exceeded further right in the heat exchanger. The zone of methane conversion therefore adapts accordingly. At low $GHSV$, a steeper gradient is obtained and the methane reaction zone shifts further to the left. If $GHSV$ is kept constant and different cases of T_{in} are assumed, the same axial shifting of

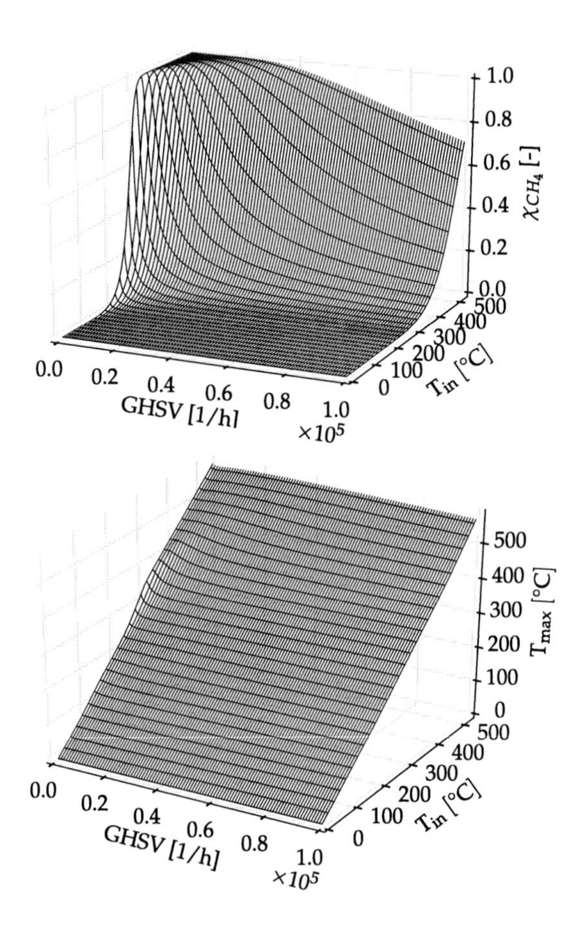

Figure 3.3.: Methane conversion (top) and maximum temperature (bottom) over *GHSV* and T_{in} under relevant operating conditions for a standard monolith.

reaction zones takes place (see Figure 2.6). However, in this case the temperature profile's sinistral gradient remains constant (see also Equation 2.35 and [49, 52, 53]), leading to almost equal maximum temperatures for all three cases in spite of T_{in} varying over 400 K.

In order to systematically study this peculiar behavior, the inflow temperature T_{in} was used as free parameter for continuations of a fully extinguished state obtained with the quasihomogeneous dimensionless model from Section 2.2.2.1. As geometric setup, a metallic countercurrent heat exchanger with 385 cpsi cell density of 0.3 m total length, wall thickness of 0.11 mm and spacer material thickness of 0.05 mm was chosen. As for the previously shown standard case, an adiabatic temperature rise of 60 K was assumed with a distribution factor Ψ of 0.2 (i.e. 12 K by CO conversion and 48 K by methane conversion). The resulting plots of maximum temperature and conversion of CO and methane are shown in Figure 3.4.

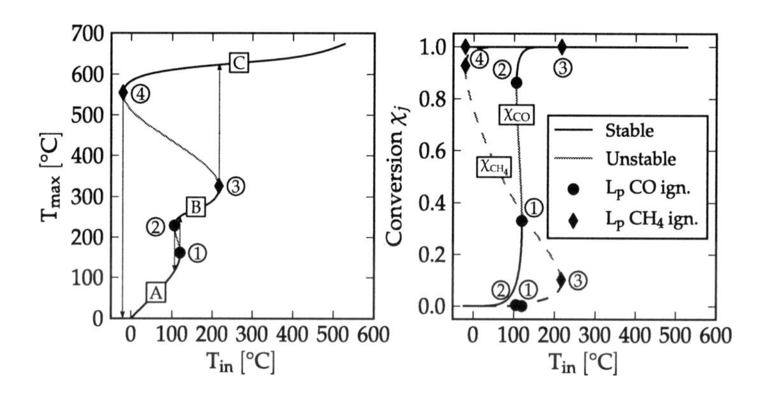

Figure 3.4.: Maximum temperature (left) and conversion of CO and methane (right) in a fully coated countercurrent reactor at constant space velocity ($GHSV = 10000\,h^{-1}$).

Compared to the standard honeycomb case, the system behaves differently due to the countercurrent heat exchange, although the same operating conditions regarding space velocity and adiabatic temperature rise were assumed. Starting from a fully extinguished state, the system is heated with increasing feed temperature until reaching point "1". While methane conversion is still very low, CO conversion has risen to about 30%. In case of further heating, the system loses stability in the singular point "1" and jumps to the intermediate stable branch (region "B" in Figure 3.4) with the CO conversion well above 90%. If the feed temperature decreases again, the system falls back to the fully extinguished state at point "2". Hence, between the two limit points "1" and "2", a first hysteresis is formed representing the ignition and extinction of CO conversion. Contrary to the standard honeycomb case, the temperature difference in maximum temperature between ignition and extinction is significantly higher which is due to the amplification factor caused by countercurrent heat exchange. If the feed temperature is increased starting from region "B", the system loses stability for a second time at limit point "3" with the methane conversion rising from about 10% to the upper stable region "C" where full conversion of both components is obtained. As can be seen in the temperature and conversion plots, this branch is very robust since both reactions can not be extinguished by a moderate decrease of feed temperature. At the same time the maximum temperature remains almost constant over a wide range of inflow temperatures, indicating the previously mentioned self-adaptation behavior of the reaction zones. Compared to the standard converter (see Figure 3.1), optimal catalyst temperatures with full conversion of both pollutants can be reached with feed temperatures above 200°C (i.e. limit point "3") which is well below the ignition temperature and T_{50} values calculated for the standard case. The fact that the ignition temperature of the feed mixture is lower than the ignition temperature of pure methane is a consequence of the admixture of CO with an ignition temperature of about 200°C (see Figure 3.2). The heat generated by CO combustion is retained in the system through countercurrent heat exchange and lowers the ignition point of the mixture.

This phenomenon is explained in the following in detail by looking at the amplification effect of the heat exchanger and the shifting reaction zones. In Figure 3.5, top left, the evolution of the amplification factor a_F over feed temperature is shown. For calculation of a_F, the maximum temperature rise obtained in the reactor is referred to the total adiabatic temperature rise of 60 K. Obviously, on the intermediate stable branch between "2" and "3", a_F is relatively small in spite of full conversion of CO since methane combustion has not yet started. The small decrease of a_F between "2" and "3" originates from the shifting of the CO conversion zone towards the reactor inlet due to self-adaptation. This leads to a decrease of the effective heat exchanger length in front of the reaction zone (see curves "2" and "3" in Figure 3.5, bottom left). Point "4" represents a fully ignited state with minimal inflow temperature. As a result, a_F is maximal with the reaction zone of methane conversion shifted to the rightmost position (curve "4" in Figure 3.5, bottom right). The maximum temperature occurs at the point where full conversion of methane is attained which is in accordance with the schematic profiles shown in [37]. Conversion of CO occurs in a zone located further left due to the lower ignition temperature. If now the feed temperature increases, both reaction zones move left. This is due to the fact that the conversion starts approximately at the point where T^* exceeds the ignition tempera-

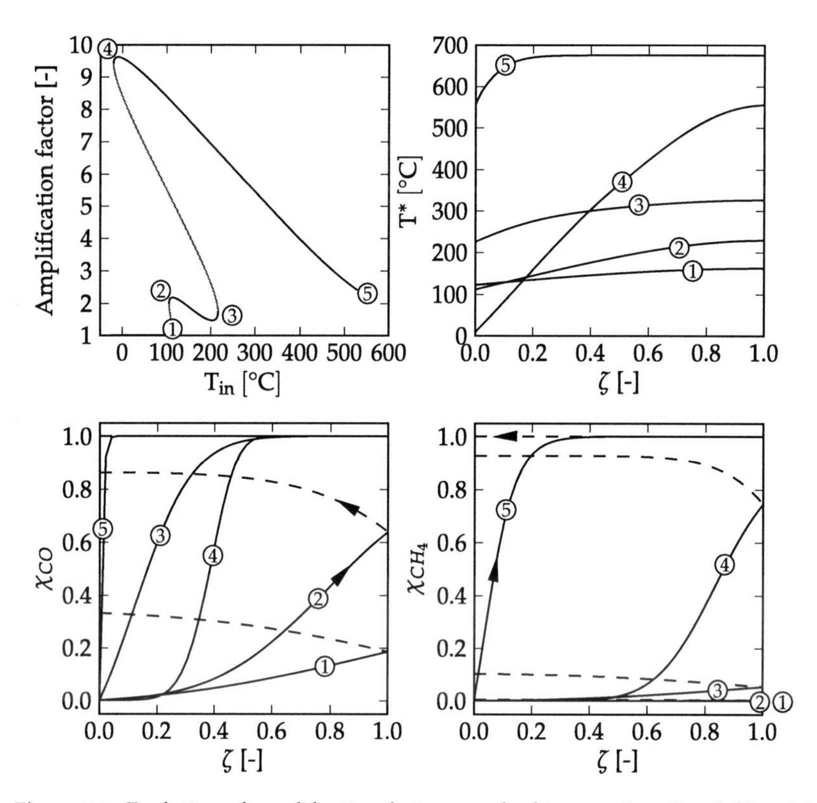

Figure 3.5.: Evolution of amplification factor over feed temperature (top left); axial temperature over dimensionless length at indicated points on the continuation curve (top right); bottom row: conversion of CO and methane obtained in inflow channels (solid line) and outflow channels (dashed line) over dimensionless length for the fully coated countercurrent reactor ($GHSV = 10000\,h^{-1}$).

ture of the respective component. Since the slope of the axial temperature profile in the heat exchange section (without reaction) does not depend on the inflow temperature (see Equation 2.35 and [49, 52, 53]), the maximum temperature remains almost constant. Consequently, increasing feed temperature causes a decrease of amplification factor.

In order to globally evaluate the performance of this system, the same procedure as for the standard case was applied (see Figure 3.3). The resulting maps are depicted in Figure 3.6. As already mentioned in the discussion of Figure 3.4 and Figure 3.5 for constant $GHSV$, unstable regions appear in the solution maps. They arise from the subsequent ignition and extinction of methane and CO, which, in combination with shifting reaction zones (see Figure 3.5), leads to significant changes of system behavior. At the limit points of these unstable regions the solution becomes singular which is characterized by a rank loss of one in the Jacobian matrix. By special reduction and projection methods, a scalar equation with local equivalence can be found [28, 39]. Right at the singularity the value of this additional function is 0. Hence, if this condition (i.e. root of additional equation) is added to the original system the rank loss is fixed and the locus of the singularity can be tracked directly as a function of two parameters ($GHSV$ and T_{in}), yielding the thick solid lines in Figure 3.6. By orthographic projection of these curves onto the parameter plane, two cusps are obtained which mark the system's stability boundaries during ignition of CO and methane (gray thick curves in Figure 3.6). The upper boundary lines hereby correspond to a continuation of the ignition limit points while the lower lines indicate the extinction. In the cusp's apex, both lines collapse into one singular hysteresis point where again the rank of the Jacobian matrix decreases by one. Similarly, corresponding test functions of higher order with local correspondance to the singular point can be found and appended to the equation system. For tracking of this hysteresis point three free parameters are required. In DIANA up to 4 parameters can be varied simultaneously in order to track singularities up to codimension 3. Mathematically, even higher-dimensional continuation would be possible. However, with each new codimension the symbolic differentiation, required for formulation of the additional test functions, becomes more difficult. Moreover, the clarity of these high-dimensional continuations from a practical point of view is usually rather doubtful.

Regarding the system's performance under conditions of low inflow temperature and space velocity, a clear benefit compared to the standard system (see Figure 3.3) can be obtained. Due to the self-adaptation of reaction zones and countercurrent heat exchange, full conversion of methane can be sustained over the complete range of inflow temperatures below 10000 h^{-1}. In case of high temperatures and throughput, the reaction zones move to the front of the device which virtually disables countercurrent heat exchange. The results therefore pass into the solution of the standard system.

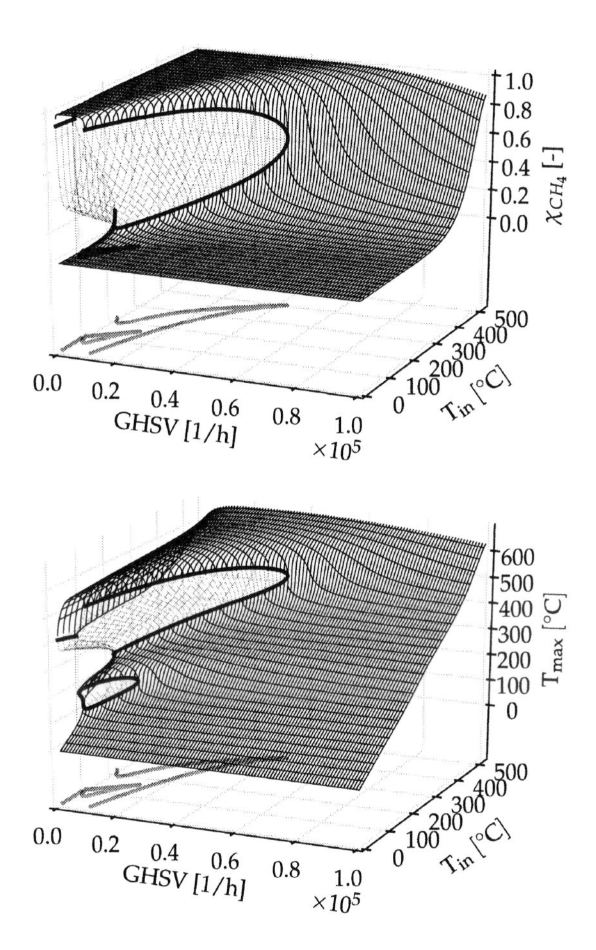

Figure 3.6.: Methane conversion (top) and maximum temperature over relevant range of *GHSV* and T_{in} for fully coated system with 385 cpsi. Further assumptions: $\Delta T_{ad} = 60$ K, $\Psi = 0.2$.

3.2.3. Partially coated heat-exchanger concept

In previous works, the concept of an integrated countercurrent heat exchanger with partially coated walls was successfully tested for different purification scenarios and with different reactor configurations [5, 24, 25, 54, 66]. This concept allows for very compact integration of the catalyst and ensures a certain amplification of the heat generated by the catalytic reactions (see also Section 2.2.2.2 for a derivation of the amplification factor). As a result, the catalyst temperature reaches much higher levels than in the completely coated case since the reaction zone is "trapped" within the coated part and self-adaptation can occur only to a limited extent.

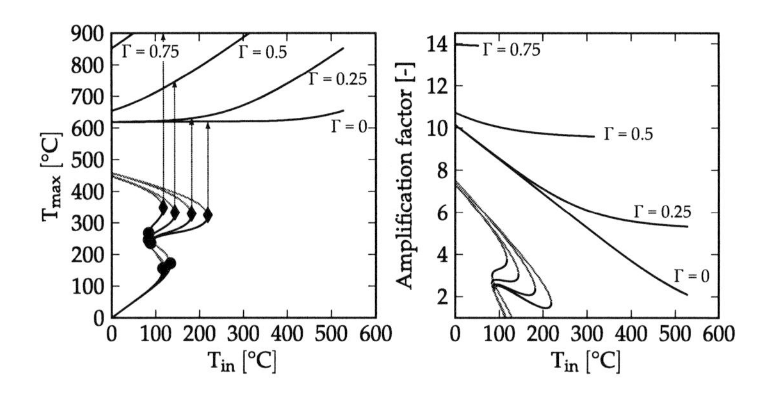

Figure 3.7.: Maximum temperature (left) and amplification factor (right) over inflow temperature for different values of uncoated heat exchanger length ($\Gamma = 0.25$, $\Gamma = 0.5$ and $\Gamma = 0.75$) compared with previously shown fully coated case ($\Gamma = 0$). Further assumptions: $\Delta T_{ad,tot} = 60$ K, $\Psi = 0.2$ and $GHSV = 10000\,h^{-1}$).

In Figure 3.7, the resulting profiles of maximum temperature and amplification factor obtained with three different values of inert heat exchanger length (see definition of parameter Γ in Section 2.2.2.2 for more details) are compared with the previously described fully coated heat exchanger ($\Gamma = 0$). All geometric parameters apart from Γ were kept equal. In order to avoid excessively high catalyst temperatures, the continuation was stopped as soon as the maximum temperature reached 900°C. The shorter the coated zone, the higher maximum temperatures are obtained which is in accordance with Equation 2.37. In fact, the amplification factor a_F (Figure 3.7, right) depends linearly on Γ, as stated by Equation 2.38 and shown graphically in [37]. The physical background of this behavior is related to the effect of sliding reaction zones described in the preceding Section (see Figure 3.5). In case of smaller coating length the reaction zone is not able to migrate towards reactor inlet for increasing inflow temperatures as observed in the fully coated case. Hence, after ignition of CO, which occurs almost independent of Γ at inflow temperatures similar to those observed in case of a standard converter (see Figure 3.2), the self-adaptation ability is increasingly hindered by shortening the coated zone. This is especially evident for $\Gamma = 0.75$ where methane conversion kicks in at values well below the ones obtained with the fully coated system or the standard approach due to the increased temperature level on the intermediate stable branch. After ignition of methane and further increasing T_{in}, the maximum temperature remains constant as long as the reaction zone is able to shift. During this phase the amplification factor decreases as observed in case of the fully coated system in Figure 3.5. This decrease stops as soon as the reaction zone reaches the beginning of the coated part, causing the maximum temperature to increase alongside with T_{in}. From that point on the amplification factor remains constant.

As for the fully coated case, the system's operational behavior in the complete range of relevant operating conditions was tested for the case with $\Gamma = 0.75$. The resulting map is depicted in Figure 3.8.

Evidently, this approach outmatches the fully coated system regarding operational stability under conditions of low temperature and throughput in spite of the considerably reduced catalyst volume. The ignition boundary of methane is shifted towards lower temperatures and the region in which full conversion of methane can be obtained is significantly broadened. However, a zone of excessively high maximum temperatures ($> 900°C$) emerges in this part of the map due to the increased amplification factor (see Figure 3.7 for comparison). Therefore, in a realistic system appropriate operating strategies are required in order to keep the temperature within the defined bounds (see Chapter 4 for a detailed description of the developed transient operating strategies). Under high throughput conditions, the shorter coated zone is slightly disadvantageous due to the reduced residence time.

3.3. Continuation of design specifications

After having discussed the characteristic behavior of three different model systems based on a continuation of operating parameters, in this section a similar technique is applied in order to directly deduce design guidelines in a very condensed form. To this end, maps are generated showing the boundaries within whom the respective system fulfills specified targets, such as a certain value of methane conversion. As in

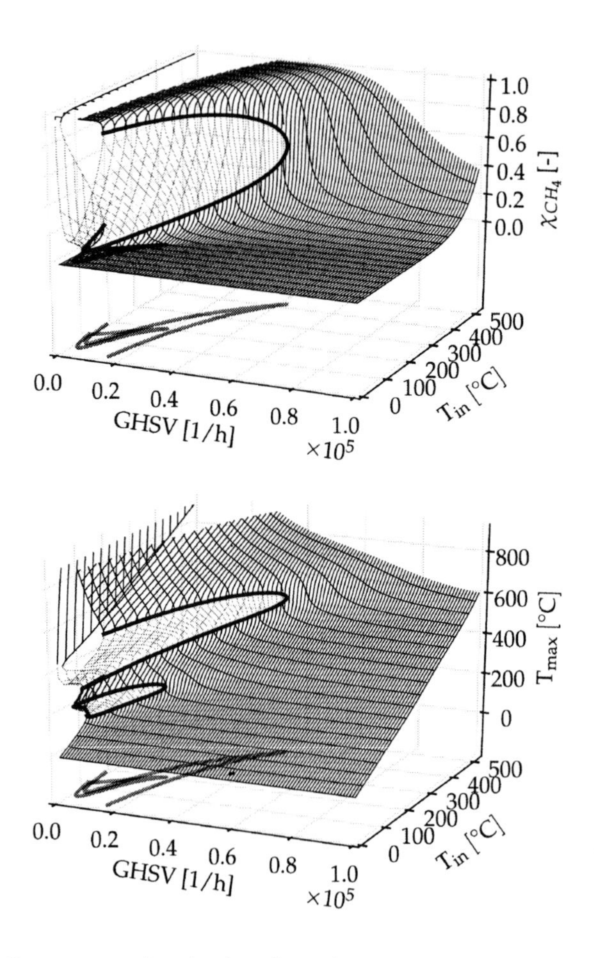

Figure 3.8.: Methane conversion (top) and maximum temperature over relevant range of *GHSV* and T_{in} for partially coated system with 385 cpsi and $\Gamma = 0.75$. Further assumptions: $\Delta T_{ad} = 60$ K, $\Psi = 0.2$.

the previous section, these results are used to underline the beneficial impact of heat integration on operational robustness.

3.3.1. Standard ceramic honeycomb

As in the previous section, the standard system comprising a catalytically coated ceramic honeycomb is used as benchmark example. While in Figure 3.1 and Figure 3.3 the stationary conversion performance over the complete range of inflow temperatures and space velocities is shown, a certain conversion value of methane is enforced in this case. One of the two operating parameters has to be declared as state variable in order to keep the equation system balanced. If now the initial stationary solution is continued as a function of the other operating parameter, the obtained curve represents the boundary line in the solution space at which the formulated specification is just fulfilled. Hence, instead of calculating the complete three-dimensional solution space and interpolating the curve of equal methane conversion, it can be directly calculated.

From the balance equations derived in section 2.2.2.4 it becomes quite obvious that the most crucial design parameter is the specific surface for heat and mass transfer a_v, since it shows up in the gas phase mass balance (Equation 2.43) and in the relation for NTU (Equation B.39). Moreover, a_v directly influences the hydraulic diameter of the channels and the gas void fraction ϵ. The effect of total length is expected to be rather negligible since it only appears explicitly in the relation for Pe (Equation B.38) and is otherwise eliminated by expressing the equations in terms of the volume-based space velocity τ. In the following the operating map of the standard system calculated with the previously described method is discussed. In Figure 3.9 curves of specified conversion values in the system's operating range are depicted. In fact, this plot represents a projection of lines in Figure 3.3, where the specified conversion value is reached, onto the operating parameter plane. Each line was directly calculated based on a continuation of a consistent initial state with a certain design configuration. In the upper row, a_v is kept constant while for each calculation run a different value of total length is assumed. As initially supposed, this design parameter has no influence on the conversion performance if $GHSV$ is used as basis for the comparison of different setups. While methane light-off (i.e. 50% conversion) can be reached or even exceeded with sufficiently high inflow temperatures in the complete operating range (top left), full conversion above 98% is only feasible below 20000 h^{-1} (top right). If in turn different values of a_v are used and total length kept constant (bottom row), considerable improvements in methane conversion performance can be obtained. The zone of full conversion extends up to 40000 h^{-1} provided that the exhaust is sufficiently hot. However, increased a_v directly leads to increased back-pressure due to the reduced hydraulic diameter.

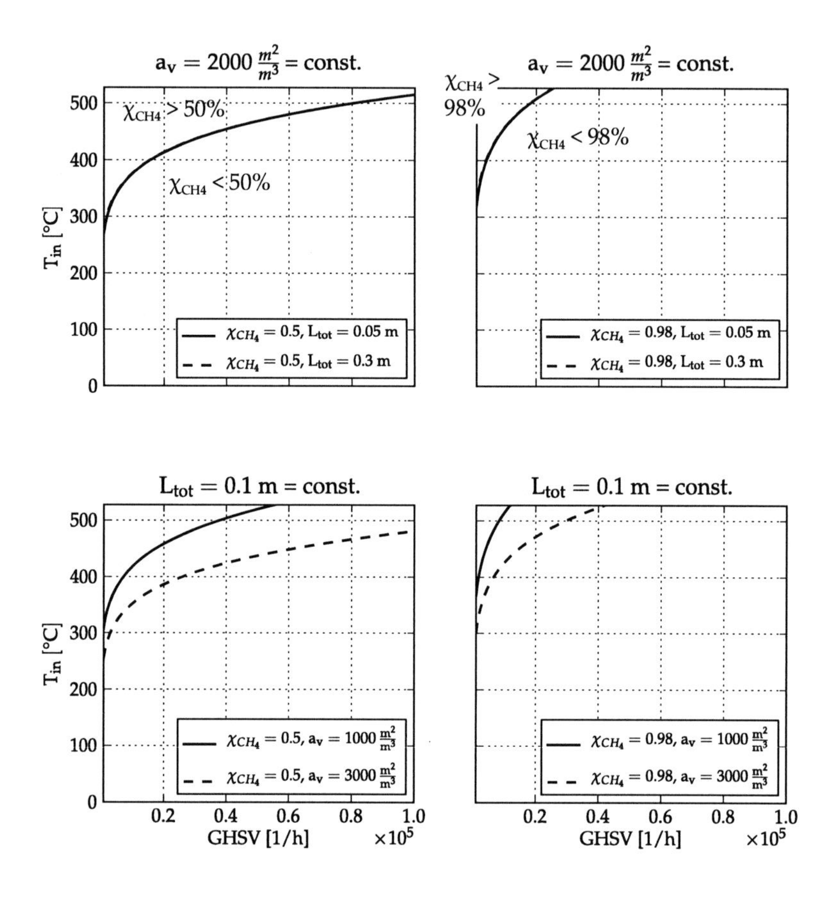

Figure 3.9.: Curves of specified CH_4 conversion values in operational parameter space for standard ceramic monolith. Top row: different cases of total length L_{tot} at constant specific surface a_v. Bottom row: different cases of a_v for constant L_{tot}. Further assumptions: $\Delta T_{ad} = 60$ K and $\Psi = 0.2$.

3.3.2. Heat-integrated systems

In the second step, the two heat-integrated concepts described in the previous section are evaluated. As for the standard ceramic monolith case, the influence of total length is expected to play a minor role. In contrast to that, the specific surface a_v enhances both heat exchanger efficiency and conversion performance. While for the standard ceramic monolith a monotonically increasing conversion plane for rising values of T_{in} was found, this is not the case for the heat-integrated systems (see Figure 3.6 and Figure 3.8). Especially at low space velocities the stable conversion plane is interrupted by unstable parts, leading to sudden jumps in conversion and maximum temperature. Consequently, the calculation of system states where 50% of methane conversion is achieved would only be feasible outside of regions where unstable states can occur. Therefore, continuations were only performed with 98% of methane conversion as target. In Figure 3.10 the resulting curves are shown for the fully coated system ($\Gamma = 0$, left column) and the partially coated system ($\Gamma = 0.75$, right column). Each case was first evaluated for different values of L_{tot} with constant a_v. In the second step, L_{tot} was kept constant and different values of a_v were tested.

Obviously, the range in which the specified conversion can be obtained can be significantly extended compared to the standard system's results depicted in Figure 3.9. For sufficient values of a_v, 98% of methane conversion is feasible over a wide range of $GHSV$. For low space velocities this result becomes even independent of the chosen inflow temperature. Due to the large catalyst volume the fully coated approach is clearly beneficial under conditions of high $GHSV$. However, for low space velocities, the partially coated concept is slightly superior due to the higher amplification factors (see also Figure 3.7 and Figure 3.8 for comparison). Since the performance under conditions of low temperatures and space velocities is more crucial in the present case of CNG exhaust aftertreatment, the partially coated countercurrent heat exchanger represents an option which allows to achieve similar or even better performance as the fully coated system with significantly less washcoat material. The overall system performance is in any case strongly influenced by the specific surface a_v, which is responsible for heat and mass exchange. High values of a_v lead however to increased backpressure which marks a severe disadvantage of heat-integrated exhaust purification systems.

Therefore, in the last step a sequential approach (see Section 2.2.2.3) is evaluated in which an inert heat exchanger is combined with a standard ceramic monolith. In order to obtain results which are comparable to the previously shown integrated cases, a system with equal ratio of heat-exchanger length to total length (i.e. $\Gamma = 0.75$) was set up. However, contrary to the integrated case, in the sequential system different values of a_v can be defined for the heat exchanger and the catalyst part, allowing to study the different impacts on the system's performance.

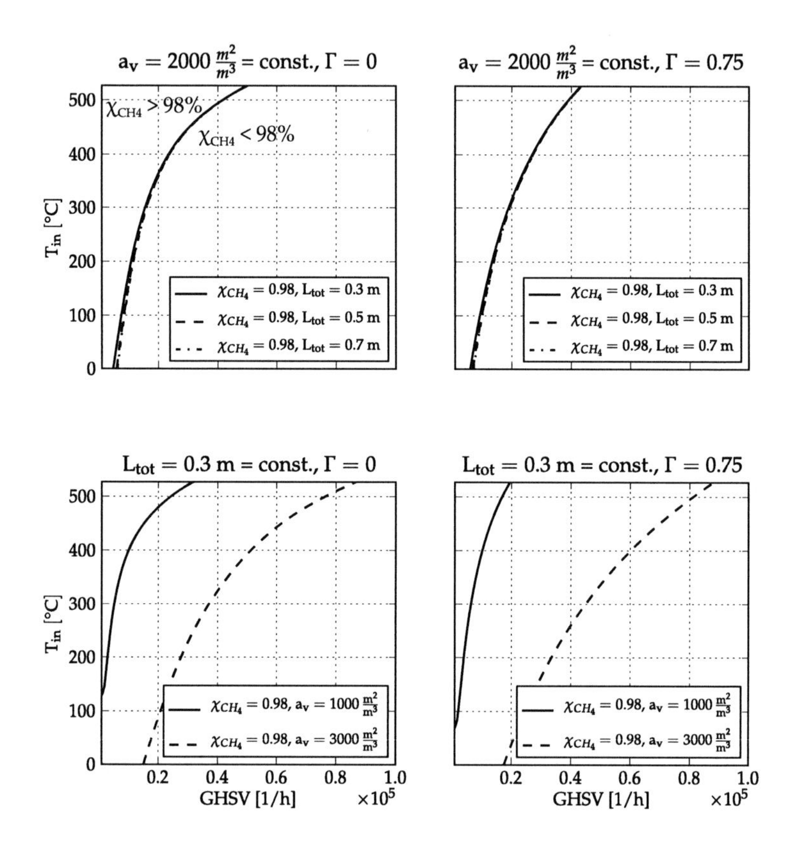

Figure 3.10.: Curves of specified CH_4 conversion values in operational parameter space for fully coated (left column) and partially coated (right column) heat-integrated systems. Top row: different cases of total length L_{tot} at constant specific surface a_v. Bottom row: different cases of a_v for constant L_{tot}. Further assumptions: $\Delta T_{ad} = 60\ K$ and $\Psi = 0.2$.

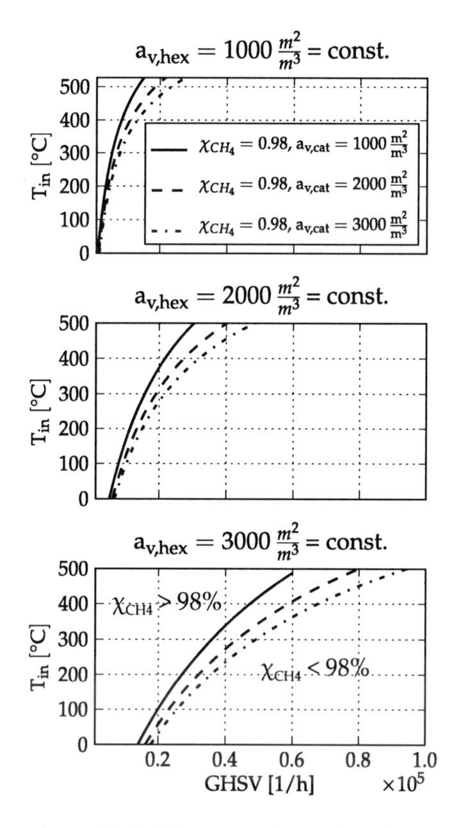

Figure 3.11.: Curves of specified CH_4 conversion values in operational parameter space for sequential system with different values of a_v in the heat exchanger. Top row: $a_{v,hex} = 1000 \frac{m^2}{m^3}$, center row: $a_{v,hex} = 2000 \frac{m^2}{m^3}$, bottom row: $a_{v,hex} = 3000 \frac{m^2}{m^3}$. Further assumptions: $\Delta T_{ad} = 60$ K, $\Psi = 0.2$ and $\Gamma = 0.75$.

The resulting plots for three different values of a_v in the heat exchanger and in the catalyst are depicted in Figure 3.11. Obviously, $a_{v,cat}$ hardly affects the system behavior for low values of *GHSV* but can help to significantly improve the conversion performance under conditions of high throughput. This is plausible since for low inflow temperatures the reactions are kinetically limited. Increased values of $a_{v,hex}$, however, lead to higher amplification factors and elevated catalyst temperatures. At high space velocities the amplification factor decreases, which is why under these conditions increased catalyst surface becomes more important. In conclusion, $a_{v,hex}$ and $a_{v,cat}$ complement one another along the continuation curve. Hence, it is not possible to fully compensate low $a_{v,hex}$ with high $a_{v,cat}$. The required values therefore have to be determined based on the available exhaust temperature level and the respective conversion target.

3.4. Conclusions

The parameter continuation tool, which is part of the DIANA simulation environment, allows to efficiently track the stationary solution of an equation system over large operating parameter ranges. This technique was applied to illustrate the characteristic differences in operating behavior between a standard ceramic catalyst and heat-integrated systems. As chemical reactions, the oxidation of CO (low ignition temperature) and methane (high ignition temperature) were considered since these components contribute most of the heat released by catalytic pollutant conversion in the exhaust of CNG powered vehicles.

While in the standard system full conversion of methane can only be achieved with relatively high feed temperatures and limited space velocity, heat integration leads to considerably improved conversion performance under these conditions. This is due to the heat amplification effect caused by the countercurrent heat exchanger which allows for significantly increased catalyst temperatures. With limited coating length, the heat amplification effect was found to be even stronger under conditions of low space velocity and feed temperature. However, the decreased catalyst volume impinges on conversion performance at high space velocities.

In order to find the best compromise between low backpressure and high conversion of methane in the relevant range of operating conditions, a sequential heat-integrated setup with separate catalyst brick and inert heat exchanger was finally evaluated. It turned out, that increased cell density in the catalyst part influences conversion performance especially at high space velocities. Increased cell density in the heat exchanger leads to an overall improvement of performance due to higher amplification factors. However, the gain in heat exchanger efficiency is accompanied by increased backpressure. Hence, with the sequential approach the heat-integrated system can be specifically adapted to the respective requirements and boundary conditions. In the following chapters, it will be shown that this design approach is also beneficial for transient operation in a vehicle.

Chapter 4.

Dynamic simulations

As demonstrated in Chapter 3, heat-integrated systems usually exhibit very robust operational characteristics if a stable ignited state is assumed as initial condition. This is due to a self-adaptation effect of the system in which the reaction zone shifts axially depending on the respective feed temperature and length of the coated zone. As a direct consequence, the maximum temperature remains in an optimal range for full conversion of methane (see Figure 3.4). In fact, with the system dimensions and operating conditions shown in Section 3.2.2 it is not possible at all to extinguish catalytic conversion of methane by a decrease of feed temperature down to 0°C. The amplification effect of the heat exchanger is in any case sufficient to maintain the catalyst temperature above light-off.

While there is a clear benefit under stationary conditions, transient operation of heat-integrated systems is challenging due to their slow response to any thermal perturbation. In a state-of-the-art approach, catalytic conversion can be quickly initiated by increasing the exhaust temperature above the required light-off level. Axial propagation of the temperature front into the monolith then occurs predominantly due to convective heat transport by the hot gas flow. However, opposite flow direction between inflow and outflow channels of a countercurrent reactor neutralizes convective transport. Hence, propagation of thermal perturbations can occur only by heat conduction in the wall material. Evidently, compared to convective transport this mechanism plays a minor role which is expressed by the relatively large values of Pe (i.e. the ratio between heat transport by convection in the gas flow and by conduction in the solid). Even more severe is the situation in case of a partially coated setup where the catalytic part is located opposite of the inflow end. As a result, catalyst heating is delayed by the time required for the thermal front to penetrate the uncoated part of the heat exchanger before actually affecting the catalyst temperature.

This chapter demonstrates the unacceptably long heating times resulting from this behavior. As a consequence, a special bypass/flap strategy is introduced in which the exhaust directly enters at the coated part and exits through the outflow channels of the converter during cold start. Thereby, the beneficial effect of convective heat transport can be exploited for catalyst heating since only the outflow channels are passed by the exhaust. After sufficient heating, the flap is closed and the system is operated in countercurrent mode. For optimization of the flap's operating strategy, simulation results with raw emission data of a CNG-fueled vehicle performing one NEDC on a chassis dynamometer are used. As exhaust aftertreatment system, an appropriately scaled partially coated heat exchanger is assumed.

In the last step, the newly developed sequential setup comprising an inert metallic heat exchanger and a catalytically coated ceramic monolith is presented. Comparative NEDC simulation results of both the sequential system and the previous integrated approach are shown in order to underline the beneficial effects of the new concept especially during catalyst heating.

4.1. Transient behavior of partially coated heat exchanger

4.1.1. Heating in countercurrent operating mode

In order to demonstrate the slow transition of a partially coated countercurrent heat exchanger from ambient temperatures to a stable ignite d state during heating, dynamic simulations with the detailed model (see Section 2.1.1) were performed. As stationary limit, which is eventually reached during heating, the solution of a quasi-homogeneous model case according to Section 3.2.3 was continued, assuming appropriate parameters (i.e. $\Gamma = 0.75$, $\Delta T_{ad,tot} = 30\ K$, $\Psi = 0.2$ and $GHSV = 10000\ h^{-1}$). The feed temperature was chosen as free parameter. Compared to Section 3.2.3, the adiabatic temperature rise is lower in this case in order to avoid excessive maximum temperatures. For the detailed dynamic model the same geometric and operating parameters were applied. Three cases with different constant feed temperatures were considered with a cold system (i.e. 20°C) as initial condition. The feed temperatures were chosen such that finally an ignited steady state with full conversion of CO and CH_4 was reached. In Figure 4.1, top, the continuation curve of the steady-state solution is shown (gray line) together with waypoints illustrating the evolution of average catalyst temperature over time. In the bottom picture, detailed heating trajectories for the three different feed temperature cases are shown.

Evidently, increased feed temperature facilitates axial expansion of the thermal front. As a result, the respective ignition levels of CO and CH_4 (i.e. $L_{p,ign.\ CO}$ and $L_{p,ign.\ CH_4}$) are reached the sooner the higher feed temperatures are applied. For instance, in case of the lowest feed temperature assumed (i.e. 200°C), it takes about two hours for the system to reach temperatures sufficiently high for CH_4 conversion. Reaching this level, which is actually well above the inlet temperature, is only

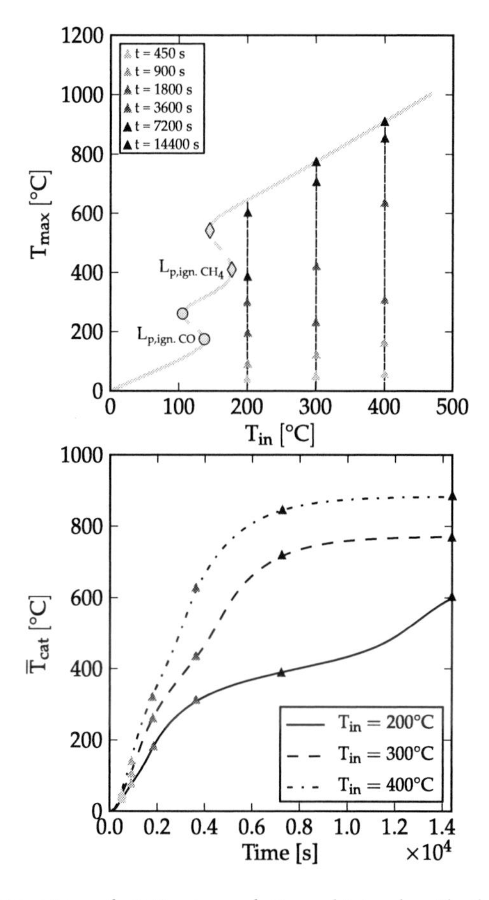

Figure 4.1.: Continuation of stationary solution obtained with the quasihomogeneous model assuming $\Gamma = 0.75$, $\Delta T_{ad,tot} = 30\ K$, $\Psi = 0.4$ and $GHSV = 10000\ h^{-1}$ (top, gray line). Average catalyst temperature at discrete points during heating (top, markers) and as continuous curve (bottom) calculated with detailed model.

possible by amplification of the heat released by CO conversion. If the feed temperature is increased to 300°C, the heating time till CH_4 ignition is halved to one hour. In case of the highest feed temperature, methane conversion is already complete after one hour. In order to further elucidate the transient development of axial temperature and conversion profiles, snapshots of the axial temperature and conversion profiles were taken at the respective waypoints shown in Figure 4.1. The resulting plots are depicted in Figure 4.2. Each column represents one feed temperature with the axial plots of wall temperature in the top row and the ones for conversion of CO and CH_4 below. Obviously, in all of the three cases shown, relatively long initial heating is required until the temperature in the coated part begins to rise (450 s). After additional 450 seconds, CO conversion starts and significantly promotes catalyst heating. Yet, even at $T_{in} = 400°C$ the adiabatic temperature rise of this first reaction (i.e. 12 K) is not sufficient to immediately reach the level required for ignition of methane. Therefore, additional heating time is required during which the catalyst temperature slowly increases due to the amplification effect of the countercurrent heat exchanger. Finally, CH_4 ignites and the adiabatic temperature rise increases to the full level. However, as during the previous phase, relatively long operating times are required until the final stationary level is reached.

After having shown the transient evolution of axial profiles during heating in countercurrent mode the main issue becomes clearer from a phenomenological point of view. Initially, the highest temperature is reached at the inflow end and the coated zone remains cold. In the final stationary profile however, T_{max} is reached in the coated zone and the lowest temperature occurs at converter inlet. As a consequence, the main challenge for an optimized heating strategy is to support the final shape of the axial temperature profile right from the start, in order to avoid the slow transient rearrangement which can only occur by heat conduction in the solid.

4.1.2. Heating with flap/bypass system

For improved catalyst heating the heat exchanger was extended with an exhaust flap and a bypass branch according to the schematic shown in Figure 4.3. The axial orientation of the heat exchanger was reversed in order to minimize the distance between exhaust gas entrance and catalyst during heating. Initially, an ideal flap performing a transition from fully opened to fully closed at a certain switch time was assumed. During heating the flap is opened, the exhaust enters the heat exchanger at the coated part and exits through the outflow channels. Hence, with the heat exchanger operated in unidirectional configuration, the thermal front is expected to quickly migrate through the coated part similar to a reference system based on a monolith brick. As soon as the complete catalyst zone is heated to or above the ignition temperature of CO (for $T_{in} = 200 - 300°C$) or CH_4 (for $T_{in} = 400°C$), which

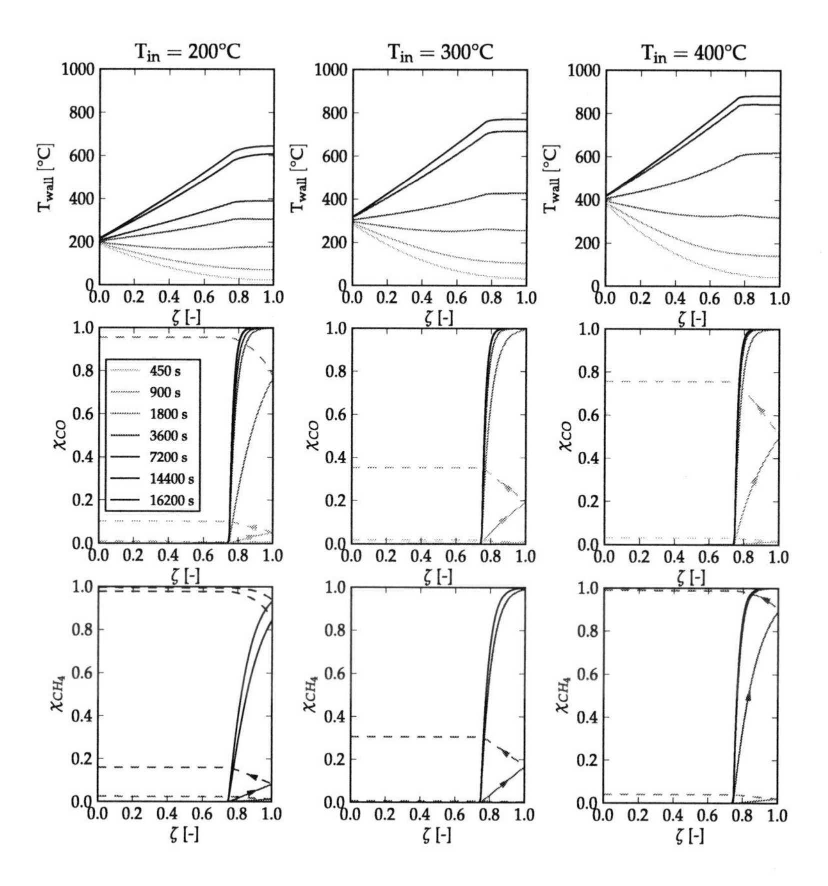

Figure 4.2.: Axial profiles of wall temperature (top row), CO conversion (center row) and CH_4 conversion (bottom row) for three different constant inflow temperatures at the waypoints specified in Figure 4.1, top.

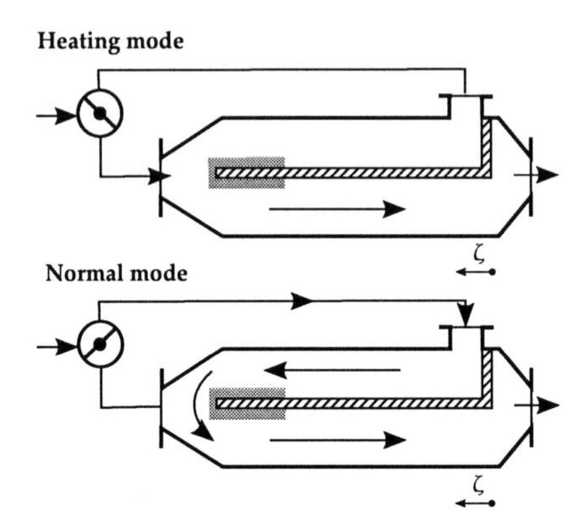

Figure 4.3.: Schematic of heat-integrated system with bypass. Cold start configuration for improved catalyst heating (top) and normal operating mode with countercurrent flow (bottom).

can be identified by the continuation solution shown in Figure 4.1 (top), the flap closes and the heat exchanger is operated in countercurrent mode with the exhaust entering at the opposite end.

In addition to catalyst heating, the flap can also be re-opened during high-load conditions when the exhaust temperature is well above the light-off range and amplification of the heat liberated by catalytic conversion of pollutants is not required. Then, the backpressure can be reduced significantly which is of special interest in order to maximise the engine's mechanical power output.

In Figure 4.4, the transient evolution of the average catalyst temperature is shown similar to the previous case without flap system (Figure 4.1, bottom). For a closer look on the initial heating period, the catalyst temperature during the first 1000 seconds is shown in a separate diagram. Obviously, the required ignition temperatures are reached much sooner than in the previous case with countercurrent heating since the hot exhaust now enters directly at the coated part of the heat exchanger. Yet, as soon as the flap is closed and the converter operates in normal mode, a pronounced decrease of catalyst temperature occurs in case of the shortest preceding heating phase (dashed line, $T_{in} = 300°C$). After about 400 seconds, a minimum is reached and the temperature increases again. For longer initial heating the decrease is weaker (solid line, $T_{in} = 200°C$) or vanishes entirely with the catalyst temperature remaining constant (dash-dotted line, $T_{in} = 400°C$). In order to study this effect in more detail, snapshots of axial profiles are taken at the waypoints defined in Figure 4.4 with special focus on the time period shortly after transition to countercurrent

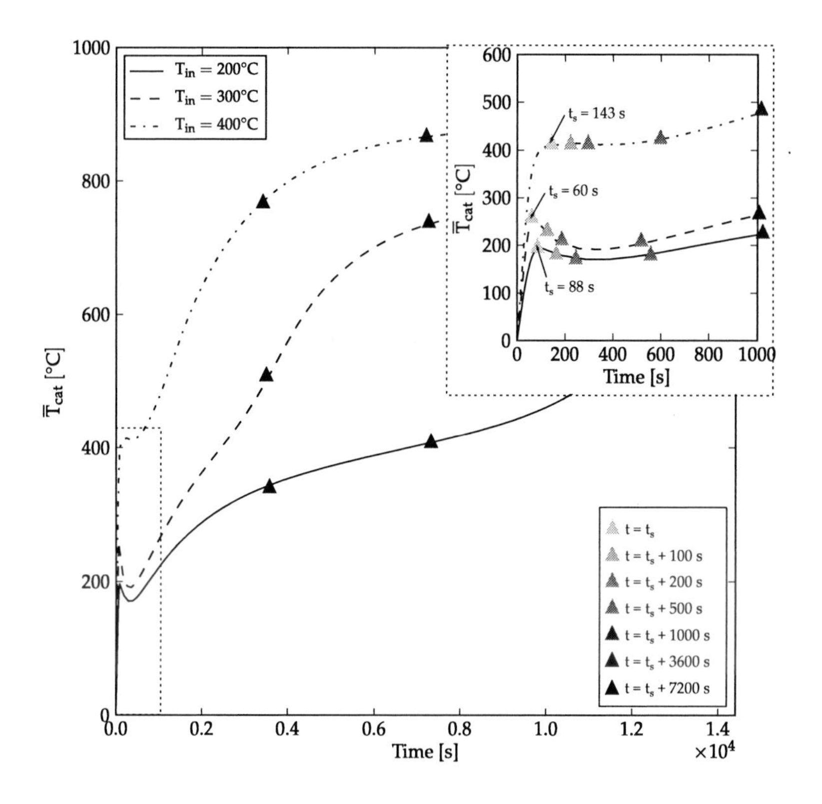

Figure 4.4.: Average catalyst temperature versus time for three different cases of constant feed temperature with initial bypass phase until exceeding $L_{p,ign.CO}$ (for $T_{in} = 200°C$, $300°C$) or $L_{p,ign.CH_4}$ (for $T_{in} = 400°C$). Insertion: initial 1000 seconds of operating time for detailed resolution of transition from bypass to countercurrent operating mode. Markers denote snapshots of axial profiles shown in Figure 4.5.

operating mode. The results are depicted in Figure 4.5. As assumed above, the axial position of the thermal front at switch time depends on the length of the initial heating period. In any case, when the bypass flap is closed CO has ignited while conversion of CH_4 is about to start only in case of the highest feed temperature. At the first profile after transition to countercurrent mode ($t_s + 100$ s), the conversion of both species is higher since now the full catalyst volume is passed by the exhaust. The temperature at converter inlet increases simultaneously due to the inflowing hot exhaust. However, since the reactor now operates in countercurrent configuration the thermal front migrates very slowly from left to right. Consequently, a zone of low temperature is formed between the preheated coated zone and the inflow end. The complete balancing of the temperature profile takes about 500 seconds in all of the three cases shown. During this time the axial profile levels out in the coated part which causes a drop of catalyst temperature and conversion in case of the lower two feed temperatures where the thermal front has not yet passed through the complete coated volume when the flap closes. Only for the highest feed temperature an almost constant level is already reached at switch time and maintained during the initial period of countercurrent operation. During the second phase of the heating procedure the system behaves similarly to the initial case with continuous countercurrent operation. The axial temperature profile remains flat in the coated part while the level slowly rises due to an increasing gradient in the uncoated part. Since adaption of the axial temperature profile to the shape of the final stationary one occurs only by heat conduction in the wall, about the same total operating time is required to reach the steady-state limit case shown in Figure 4.1.

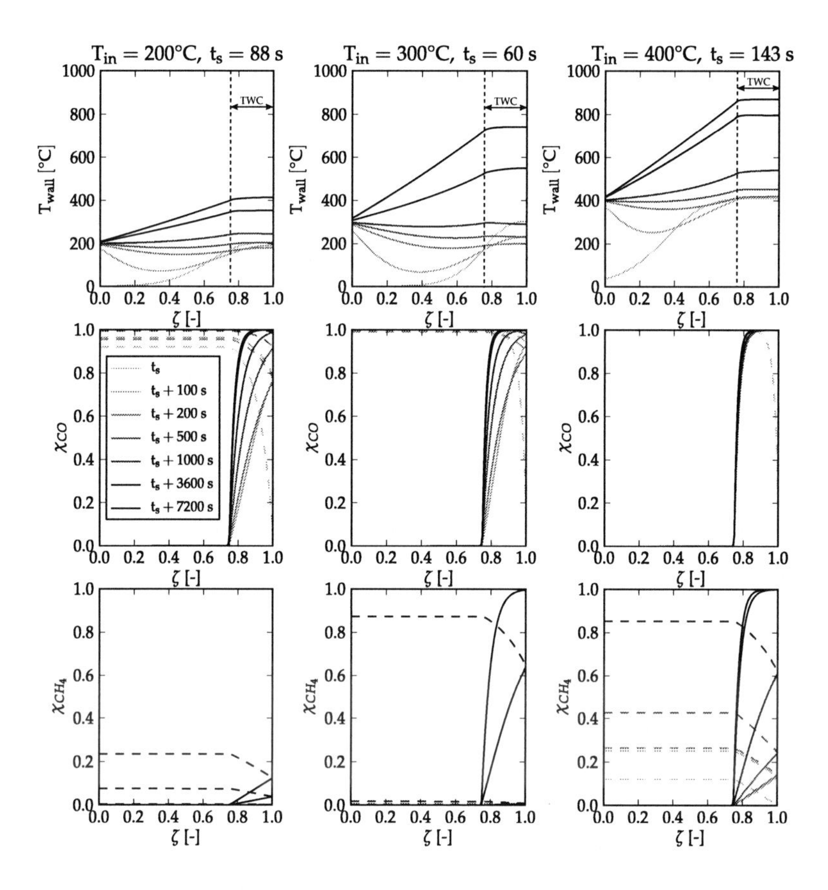

Figure 4.5.: Axial profiles of wall temperature (top row), CO conversion (center row) and CH_4 conversion (bottom row) for three different constant inflow temperatures at the waypoints specified in Figure 4.4.

4.2. Heating strategies for operation under drive cycle conditions

4.2.1. Feed conditions for drive cycle simulations

In spite of the usually significant discrepancy between measurement data obtained during normalized drive cycles and emission data collected under more realistic conditions [58, 68], the New European Drive Cycle (NEDC) is still the reference test cycle during homologation of new vehicle prototypes. In case of this simulation study, raw emission data obtained during a NEDC sequence of a CNG powered monovalent vehicle featuring a 1.8 l turbo-charged stoichiometric engine was used. For improved numerical stability and reduced simulation time, continuously differentiable cubic splines were used for interpolation in between the data points. The polynomial parameters required for these functions were calculated using the tools provided by SciPy [32]. In Figure 4.6 the resulting plots of after-turbine temperature, mass flux and pollutant concentrations (i.e. CO and CH_4) are shown. Obviously, the exhaust temperature is close to or above the level required for CH_4 light-off during most of the time. However, during idling phases especially in the urban part of the cycle (i.e. 0-800 s), the temperature level repeatedly drops well below 400°C which would lead to extinction of the CH_4 conversion reaction in a standard system without heat recovery.

4.2.2. Geometric properties of full-scale systems

As base design, the brazed heat exchanger presented in Section 5.2 was assumed. The resulting geometric parameters for the simulated full-scale system are shown in Table 4.1.

Compared to the laboratory-scale prototype presented in Chapter 5, the cross-sectional area is increased while the total length, the channel height and the sheet thickness of the wall material are kept equal. In order to decrease thermal mass, the corrugated spacer structures are assumed to be made of significantly thinner material as in case of the laboratory-scale prototype. Contrary to thinner material for the channel walls, this modification does not increase the risk of leakage formation during brazing. For minimized backpressure, the same technique as for the laboratory-scale system with low cell density in the uncoated part and increased cell density in the coated part is assumed. On the whole, the setup outlined above leads to a total coated volume of 1.6 l.

In order to evaluate the heat-integrated system, comparative results obtained with a reference case comprising a coated ceramic monolith with equal cell density and volume as the catalyst part of the heat-integrated system (i.e. 250 cpsi, 1.6 l) but with-

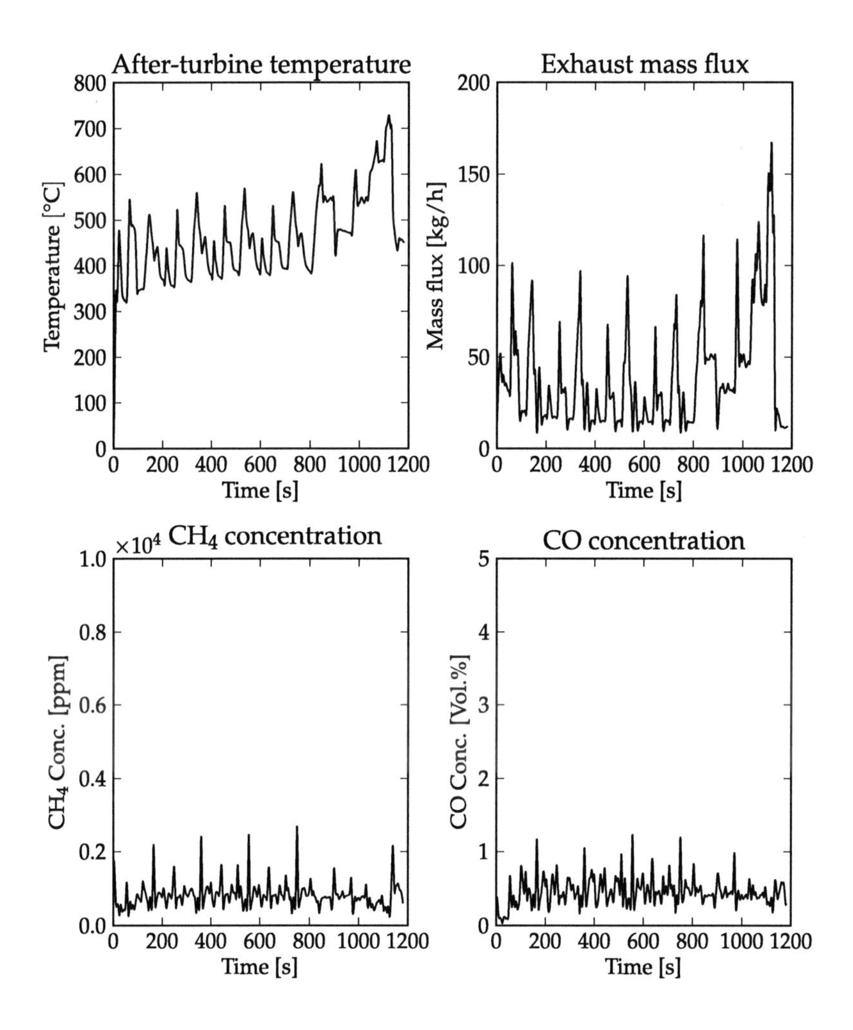

Figure 4.6.: Engine raw emissions during NEDC sequence. Top left: after-turbine exhaust temperature. Top right: exhaust mass flux. Bottom left: CH_4 concentration. Bottom right: CO concentration.

Table 4.1.: Geometric parameters of full-scale brazed heat exchanger for dynamic simulation with realistic raw emission data.

Total length	$[m]$	0.3
Coated length	$[m]$	0.1
Cross-sectional area	$[m^2]$	0.0162
Channel height	$[mm]$	2.9
Sheet thickness wall	$[\mu m]$	150
Sheet thickness spacer	$[\mu m]$	75
Cell density uncoated part	$[cpsi]$	133
Cell density coated part	$[cpsi]$	250

out heat recovery are shown. In state-of-the-art catalyst systems for CNG vehicles, the close-coupled brick is usually combined with an additional one in underfloor position. This usually leads to an increased total amount of washcoat and precious metal.

4.2.3. Bypass/flap system, no auxiliary heating

In this section, the heat exchanger setup with flap/bypass system and reversed axial orientation, as introduced in section 4.1.2, is applied. Direct attachment of the coated end to the turbo charger is assumed using the engine-out temperature depicted in Figure 4.6 as inflow condition. Certainly, this configuration is rather difficult to implement in a vehicle due to packaging constraints in the engine bay.

4.2.3.1. Optimization of bypass flap operating strategy

Since cold start and catalyst heating has to occur rapidly during the initial phase of the drive cycle, only the first two ECE sequences (i.e. 0-400 s) are considered. Contrary to the initial study of heating behavior (Section 4.1), a global heat loss coefficient of 1.5 $W/m^2/K$ referred to the outer surface of the converter is assumed for the results presented in the following. Similar values were obtained during model fitting of the laboratory-scale prototypes. These reactors were equipped with a ceramic fibre insulation (see Section 5.1.2 and Section 5.2.2 for results).

For flap operation, two different strategies are considered. In the first case, a continuous valve is assumed whose position is allowed to change between fully opened (i.e. "1") and fully closed (i.e. "0"). The second option is represented by a simple exhaust flap, as introduced in Section 4.1.2, which can perform only one switch from opened to closed at a certain point in time during the drive cycle. In both cases, the switching strategies are optimized with a minimization of the accumulated CH_4 tailpipe emissions as objective. The optimization algorithm CONDOR [4], which is part of the DIANA optimization subsystem, was chosen as solver. In this method, the objective function is approximated by a quadratic function, thus yielding a map which depends on the parameters to be optimized. By evaluation of this map, the objective function's extrema can be predicted efficiently without the need of the original Jacobians of the model. In contrast to a previous version based on the same approximation technique [60], CONDOR also allows the formulation of constraints.

For the continuous flap approach, the transient profile of the flap position is assumed to be based on 80 discrete points equally distributed over the simulation sequence. In the intervals between, the flap position is interpolated with a continuously differentiable spline function as already applied for the feed data (see section 4.2.1). Since the solver can only handle up to 40 degrees of freedom simultaneously, the optimization is performed in two steps. At first, the flap position is optimized during the initial 200 seconds while the flap position is set to 0 in the second interval. Subsequently, the previously obtained, optimal flap profile is applied on the initial interval while the remaining 40 points (i.e. 200-400 s) are optimized.

In case of the binary flap, the only parameter for optimization is the switch time when the flap position changes from "1" to "0". As for the continuous approach, a smooth time-dependent function is applied for the flap position F, mimicking a sharp step at the respective switch time t_s:

$$F(t) = 0.5 \cdot (1 - tanh\,(t - t_s))\,. \tag{4.1}$$

Since only one parameter (i.e. t_s) has to be optimized, this approach is expected to tremendously decrease computational cost. In Figure 4.7, the optimal transient profiles of flap position are shown (top row) together with transient mass flows of CH_4 at inlet and outlet of the heat exchanger (center row) and axial temperature profiles (bottom row) for both flap operating strategies. In both cases, an open flap was found to be optimal during the initial 100 seconds. After $\sim$ 60 seconds, CH_4 conversion starts in both cases which is expressed by an increasing deviation of CH_4 outflow from the respective value at converter inlet (see Figure 4.7, center). This arises from a pronounced increase of exhaust temperature during the first acceleration phase. At the beginning of idling between 90 and 95 seconds, the continuous flap closes for the first time. As a result, the converter is operated in countercurrent mode with the exhaust passing through both the converter's inflow and outflow channels. This leads to increased conversion compared to the binary strategy where the converter is still operated with open flap. In addition to increased residence time

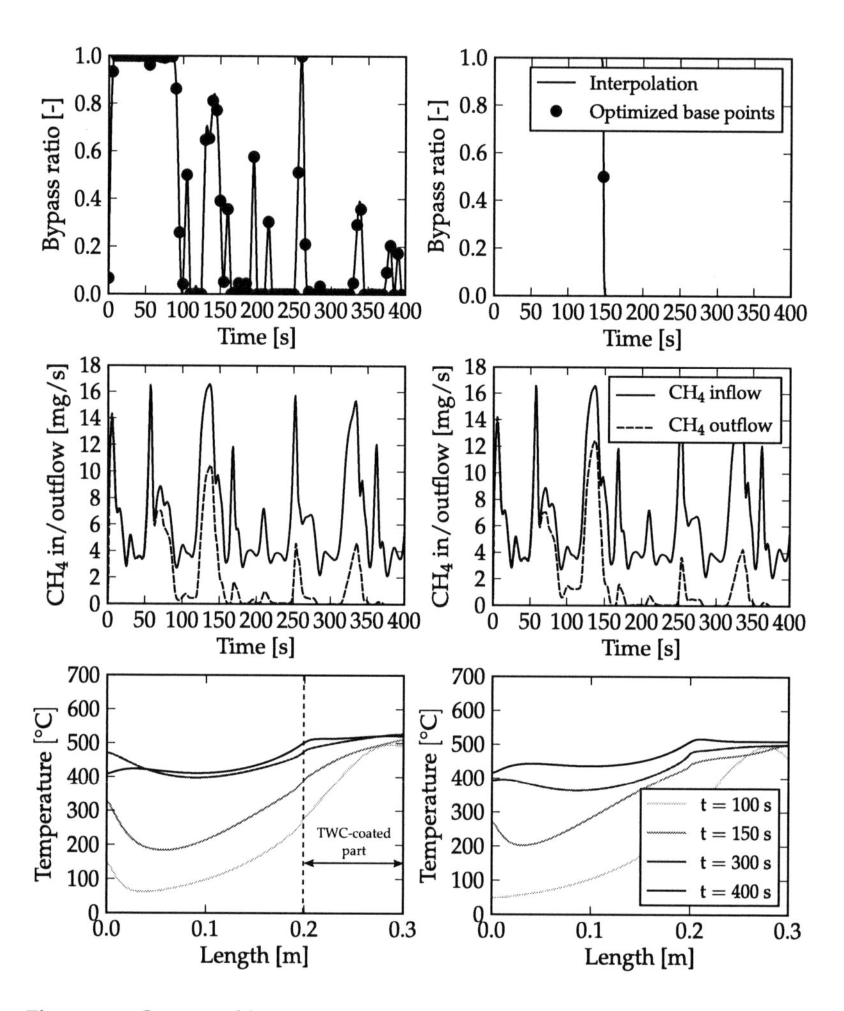

Figure 4.7.: Optimized bypass ratio of continuous (left) and binary flap (right) during initial 400 seconds of NEDC (top row). Inflowing and outflowing CH_4 (center row). Snapshots of axial temperature profiles (bottom row).

in the coated section, the catalyst is also shielded from a drop of exhaust temperature during the idling phase. This becomes obvious from the axial temperature profile at $t = 100$ s (Figure 4.7, bottom). While the temperature is maximal at the right-hand side in case of the (closed) continuous flap, it decreases in case of the binary one which remains open. As soon as the exhaust temperature rises again during the following acceleration phase, the continuous flap re-opens for further catalyst heating. Since the temperature level has been maintained during the preceding idling phase, emissions are lower during the acceleration phase in case of the continuous flap. The binary flap closes at $t = 147$ s which marks the beginning of the next idling phase. At this point (i.e. $t = 150$ s), the axial temperature profile in the catalyst section is still tilted in case of the continuous flap, leading to a lower mean temperature compared to the binary flap where the catalyst temperature is generally higher. Consequently, emissions are marginally increased during the remaining simulation in case of the continuous flap. In order to catch up on the binary flap strategy, the continuous flap re-opens during phases of high exhaust temperatures. As a result, the final axial profiles obtained at $t = 400$ s are very similar for both cases.

4.2.3.2. Emission behavior during complete NEDC - optimal bypass strategies

In this section, the benefit of the optimal, continuous flap regarding decreased global CH_4 emissions is evaluated by comparing the results obtained during one complete NEDC sequence with the ones yielded by the binary flap approach. Additionally, the simulation results of a reference system, comprising a ceramic honeycomb monolith with equal cell density and volume as the coated part of the heat exchanger (i.e. 1.6 l, 250 cpsi), are shown. As for the heat-integrated system, a global heat loss coefficient of 1.5 $W/m^2/K$ referred to the outer surface of the ceramic brick is assumed. The resulting plots of mean catalyst temperature, feed temperature and accumulated CH_4 emissions over time are depicted in Figure 4.8. Due to the low thermal mass of the reference system's ceramic monolith, catalyst heating occurs significantly faster than in case of the integrated system. Moreover, since in the latter case the exhaust only passes through the outflow channels of the heat exchanger during bypass operation, the active surface is reduced by a factor of about two compared to the reference case. Conversion of CH_4 therefore starts earlier in case of the reference system which reduces the mass of accumulated methane emissions during the initial phase. However, as soon as the heat-integrated system is operated in countercurrent mode the catalyst temperature can be maintained and even increases significantly during the remaining cycle. At 600 seconds, the catalyst temperature exceeds 550°C, thus leading to complete suppression of further methane emissions. Contrary to that, the temperature of the reference system's catalyst brick follows the feed temperature much more closely. As a result, the temperature cycles between 400°C and 500°C for

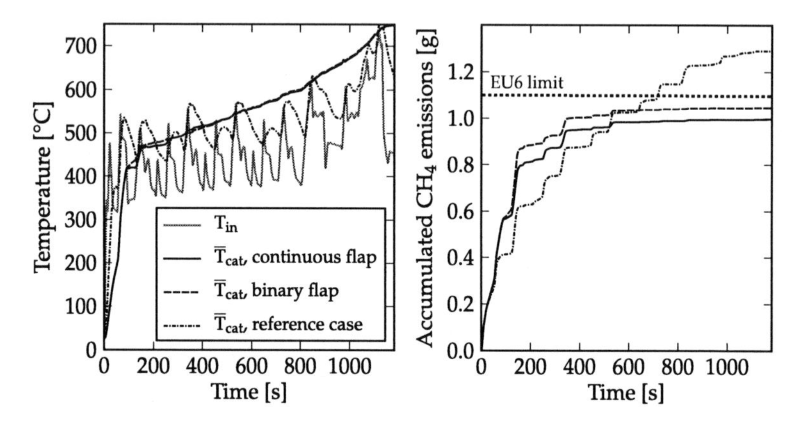

Figure 4.8.: Transient evolution of mean catalyst temperature (left) and accumulated CH_4 emissions during full NEDC sequence for different flap operating strategies and reference case without heat exchanger.

most of the time during the urban part of the cycle (i.e. 0-800 s). Methane therefore breaks through in each acceleration phase, leading to a steady increase of tailpipe emissions. Hence, the final amount of emitted CH_4 is higher than in case of the heat-integrated system and even exceeds the EU6 threshold. The beneficial effect of the continuous flap-operating strategy is expressed by a 5% decrease of global CH_4 emissions compared to the binary flap. Although the mean catalyst temperature is slightly higher in case of the binary flap during the initial 200 seconds, the strategy of early transition to countercurrent operation with intermittent re-opening of the flap during the following acceleration phases improves overall methane conversion. Yet, the gain in conversion efficiency is rather negligible compared to the significant increase of system complexity for a continuous bypass flap.

4.2.3.3. Derivation of temperature-based control strategies

In the previous section, emission results were shown using two different optimal control strategies for the bypass flap. These findings will be used in the following to derive two different control strategies which are based on catalyst temperature as input.

As in the optimal binary case, one switch point is assumed where the bypass switches to countercurrent operation. The obvious condition for this switch point is that the complete coated part has to be heated above the ignition temperature of methane. According to Figure 4.7, bottom right, this limit temperature can be specified as $\sim 470°C$. This start-up control strategy will be called binary control.

The second control strategy, called continuous control is based upon the optimized transient profile of the continuous flap approach (Figure 4.7). There it was shown, that re-openings of the flap are beneficial during acceleration phases, whenever the engine exhaust temperature exceeds the actual mean catalyst temperature. As soon as the exhaust temperature falls below the catalyst temperature, the flap should close. Hence, for the continuous control approach the flap will be either fully opened if $T_{in} > \overline{T}_{cat}$, or fully closed if $T_{in} < \overline{T}_{cat}$. In Figure 4.9, the transient evolution of bypass ratio, mean catalyst temperature and accumulated CH_4 emissions is shown for both cases. With the temperature threshold for flap actuation directly adopted from the optimal result, the same switch time ~ 150 s is obtained for the binary control approach and a similar amount of emitted CH_4 is obtained. The result yielded with the continuous control approach is also very close to the optimal one since flap actuation occurs at the same points in time (see Figure 4.7, top left). As a result, under NEDC conditions both strategies are well-suited in order to minimize methane emissions. Since the benefit of the continuous control approach is rather small, only the binary control will be applied in the following. With regard to more dynamic drive cycle scenarios, which are not considered in this work, the continuous approach might however lead to significant improvements.

4.2.4. Bypass/flap system with electric auxiliary heating

As mentioned before, direct attachment of the relatively big heat exchanger unit to the engine is rather difficult due to packaging reasons. Instead, the device can be mounted in an underfloor position. During normal operation, the temperature of the catalyst zone is kept above the level required for light-off due to the heat exchanger's amplification effect. Yet, heating during cold start becomes increasingly challenging. If an engine-based strategy (i.e. delayed ignition for increased exhaust temperature) is applied, heat losses and thermal inertia of the connection tube between engine and converter in underfloor position lead to a considerable decrease of exhaust temperature at heat exchanger inlet. Therefore, direct heat input at the coated zone is much more efficient. In a previous publication about heat-integrated exhaust purification for diesel engines [5], a fuel burner was suggested as cold start support. However, in the present case of a stoichiometric CNG engine this would lead to additional difficulties since the exhaust's air-fuel ratio has to be controlled precisely for best conversion performance of the three-way catalyst. Moreover, significant slip of CH_4 occurs at the burner during start-up which impinges on tailpipe emissions. As a result, an electric heater [46, 59], directly attached to the coated end,

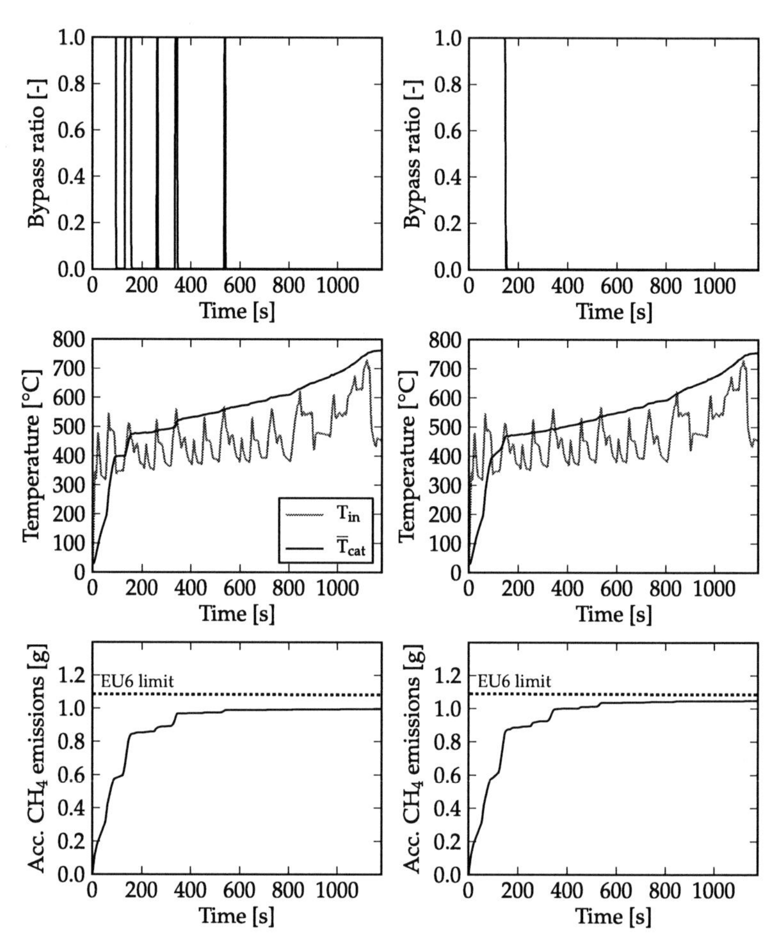

Figure 4.9.: Transient evolution of bypass ratio (top row), mean catalyst temperature (center row) and accumulated CH_4 emissions during full NEDC sequence for binary flap operating strategy (right column) and continuous approach (left column).

is better suited. In Figure 4.10, a schematic of this approach is shown. Besides the

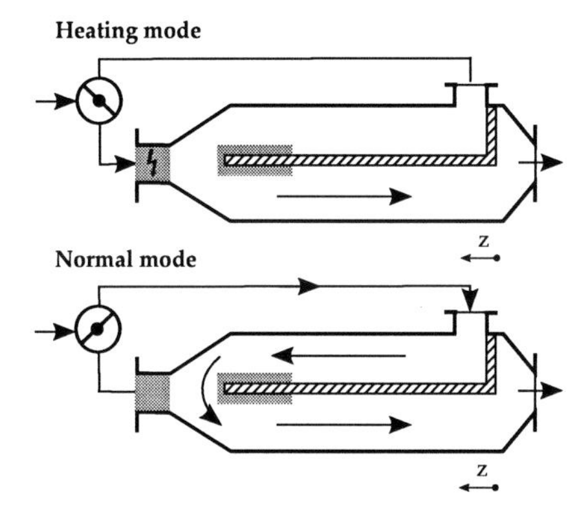

Figure 4.10.: Schematic of heat-integrated system with bypass and electric heater. Cold start configuration with active heater (top) and normal operating mode with countercurrent flow without auxiliary heating (bottom).

complete absence of secondary emissions during start-up and operation, this system is also more compact and less expensive than a fuel burner with comparable power output. For the NEDC simulations shown in the following, a simple proportional controller was chosen in order to dynamically calculate the power input required for maintaining a certain set point temperature at catalyst inlet. Hence, the heater is assumed to rapidly adapt to the respective mass flow and exhaust temperature up to a certain maximum value. It operates only during the initial bypass phase while the temperature level can be maintained during the remaining cycle due to the heat exchanger's amplification effect. For the coated part the same volume and geometric configuration as in the previous section (i.e. 1.6 l, 250 cpsi) was assumed. In order to mimic the underfloor position of the heat exchanger, the exhaust temperature was decreased considerably compared to the after-turbine temperature shown in Figure 4.6. As already pointed out in a previous publication [65], this temperature profile can be regarded as worst-case scenario. Yet, in consideration of future lean CNG engines this exhaust temperature range is likely to become increasingly realistic. In Figure 4.11 (top left), the resulting profiles of mean temperature in the coated zone of the heat exchanger are shown together with the assumed feed temperature in underfloor position. The bypass and heating period is same as for the results shown in the previous sections for the binary flap approach (147 s). The maximum heating power is adjusted such that the EU6 emission target of 0.1 g total hydrocarbon per km can be fulfilled. During heating, a set point temperature of 750°C is assumed.

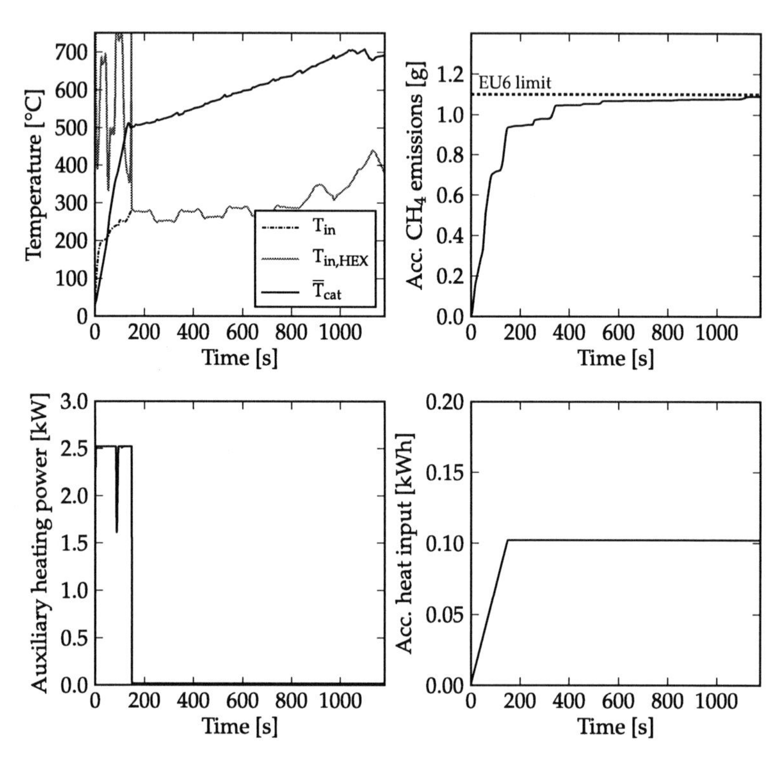

Figure 4.11.: Assumed inflow temperature and average catalyst temperature over NEDC with initial heating input (0-147 s, $T_{set} = 750°C$) (top left). Accumulated CH_4 emissions (top right). Transient evolution of heating power (bottom left) and accumulated energy input (bottom right).

This value is only reached during phases of low exhaust mass flow while the chosen maximum power output is obviously not sufficient for maintaining this temperature level during acceleration. As a result, the heater operates at maximum power for most of the time during catalyst heating. At the end of the heating phase the average catalyst temperature reaches 500°C and full conversion of CH_4 is obtained. To reach this point, 0.1 kWh of electric power are required which corresponds to about 1.2% of the vehicles total fuel consumption during the complete NEDC sequence. Concerning CH_4 tailpipe emissions, about 80% of the final value occur during catalyst heating. Hence, improving the cold start behavior of the catalyst potentially leads to an overall decrease of CH_4 emissions.

4.3. Sequential system

One possible option for decreased thermal mass and improved heating behavior is to change the design of the metallic heat exchanger in such a way that thinner material is applied for the channel walls. Yet, especially in case of high-temperature brazed heat exchangers there is usually a trade-off between risk of leakage formation by fusing of channel walls and low material thickness. Another possibility is to use ceramic monoliths as heat exchanger. In order to establish countercurrent flow, every second channel row has to be blocked and a sidewise outlet has to be provided [25]. Stationary results obtained with such a system were published previously [5] and confirmed the principal feasibility of this design approach at laboratory scale. Yet, while clear advantages can be obtained with respect to mechanical washcoat stability and thermal mass, there are also challenges and drawbacks. For instance, extrusion of long monolith bricks with thin channel walls is very challenging with respect to imperfections leading to internal bypass and decreased conversion performance. In addition to that, equal cell density has to be applied for the complete heat exchanger which inevitably leads to increased backpressure compared to a metallic system where the cell density can be changed between inert and coated part by applying different corrugated structures.

The consideration of the above mentioned assets and drawbacks finally led to the development of a sequential system comprising an inert metallic heat exchanger with relatively wide channels and a catalytically coated standard ceramic honeycomb with increased cell density [19]. This setup represents a good compromise regarding optimized cold start performance and minimized backpressure. In Figure 4.12, a sequential system with bypass flap and electric heater is schematically depicted. As in case of the integrated system, an axially reversed orientation was assumed with the coated end pointing towards the exhaust gas source. For improved cold start performance, the electric heater was directly attached to the ceramic monolith. During the heating phase the exhaust passes through the complete volume of the catalyst brick. This is a major improvement compared to the integrated system where the gas passes only through the converter's outflow channels during heating mode. In fact, heating should now occur equally fast as in case of the reference system. After leaving the catalyst, the exhaust exits through the heat exchanger's outflow channels which are thereby preheated. When the bypass flap is closed the exhaust enters at the opposite end, is preheated in the inflow channels, passes through the upper side channel and exits through catalyst brick and outflow

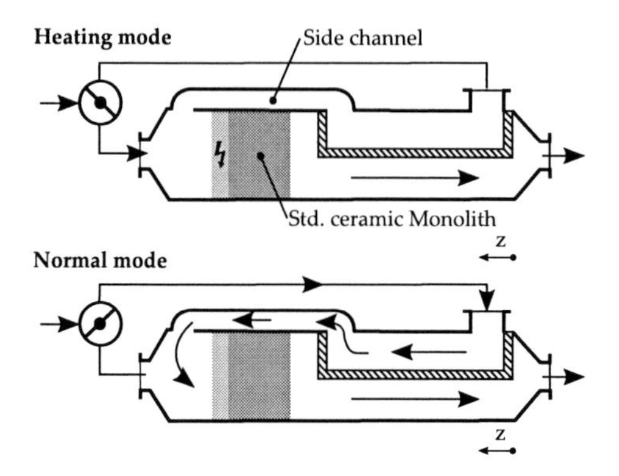

Figure 4.12.: Schematic of sequential system with bypass and integrated electric heater. Cold start configuration with active heater (top) and normal operating mode with countercurrent flow in heat exchanger (bottom).

channels of the heat exchanger. As in the integrated system, the bypass flap can be opened during conditions of high exhaust temperature in order to disable the heat exchanger's amplification effect and reduce backpressure.

4.3.1. Geometric properties of full-scale sequential prototype

As pointed out in 2.2.2.3 and confirmed by simulation results shown in 3.3.2, equal geometric layout leads to very similar stationary results with both the integrated and the sequential approach. Therefore, in order to especially evaluate the benefits of the sequential approach during cold start, a heat exchanger setup very similar to the one applied in Section 4.2.2 is assumed. However, the cell density of the catalyst part is increased, assuming a state-of-the-art substrate layout (400 cpsi, 6 mil). The resulting geometric parameters are shown in Table 4.2.

Since the cooling effect of the side channel between heat exchanger and catalyst is expected to play a major role during transient operation of the sequential system, it is incorporated into the model with a lumped simulation approach which is described in Section 4.3.2.1. Rectangular shape is assumed with appropriate dimensions to cover the distance between sidewise outlet of the heat exchanger's inflow channels and head chamber of the converter. In Figure 4.13, a sketch of the assumed channel layout is shown. While the channel height was fixed at 25 mm, the width is determined by the geometry of the heat exchanger package. The length is set to

Table 4.2.: Geometric parameters of full-scale sequential heat-integrated system for dynamic simulations with realistic raw emission data.

Total length	$[m]$	0.3
Ceramic TWC length	$[m]$	0.1
Cross-sectional area	$[m^2]$	0.0162
Channel height heat exch.	$[mm]$	2.9
Sheet thickness wall heat exch.	$[\mu m]$	150
Wall thickness TWC	$[mil]$	6
Sheet thickness spacer	$[\mu m]$	75
Cell density heat exch.	$[cpsi]$	133
Cell density TWC	$[cpsi]$	400

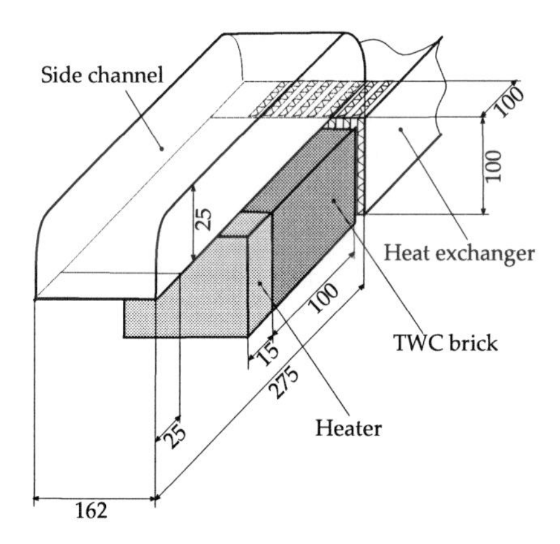

Figure 4.13.: Schematic (not drawn to scale) of side channel with dimensions according to the converter layout defined in Table 4.2.

a sufficiently high value to allow for small gaps between catalyst brick and heat exchanger (10 mm) and behind the electric heater (25 mm).

4.3.2. Comparative study on NEDC performance

4.3.2.1. Transient simulation model of sequential system

The mathematical model applied for dynamic simulation of the sequential system comprises two major parts. In the first one an inert heat exchanger is modeled, assuming the same energy balance equations as applied for the integrated system but without energy source terms. The mass balance equations are omitted and the same gas phase concentrations are assumed at inflow and outflow. The second part comprises the ceramic catalyst which is modeled with the same balance equations as the reference system whose dynamic simulation results are presented in Section 4.2.3.2. In order to simulate the side channel which connects the heat exchanger and the catalyst (see Figure 4.13), the following set of lumped energy balances was implemented:

EB channel wall:

$$m_{sc}c_p^s\frac{dT^s}{dt} = \alpha_{sc}^{g,s}A_{sc}\left(\overline{T}^g - T^s\right) - \alpha_{sc}^{amb}A_{sc}^{ext}\left(T^s - T^{amb}\right) \tag{4.2}$$

EB gas phase:

$$\begin{aligned} \dot{m}^g\overline{c}_p^g\left(T^{in} - T^{out}\right) &= \alpha_{sc}^{g,s}A_{sc}\left(\overline{T}^g - T^s\right), \\ \text{with } \overline{T}^g &= \frac{T^{in} + T^{out}}{2}. \end{aligned} \tag{4.3}$$

Ideal coupling is assumed between the outflow end of the side channel and the catalyst (i.e. $T^{out} = T^{in}_{cat}$).

The surface of the channel walls A_{sc} results from the dimensions depicted in Figure 4.13. Assuming a sheet thickness of 1 mm for the walls, the mass m_{sc} is obtained. For the heat capacity c_p^s and density ρ^s of steel, the same constant values as for the heat exchanger are applied. The gas phase heat capacity $\overline{c}_p^g$ is calculated based on the average gas phase temperature $\overline{T}^g$. The resulting values of all parameters required for evaluation of Equation 4.2 and Equation 4.3 are shown in Table 4.3.

The heat transfer coefficient for heat losses from the side channel to the surroundings is significantly higher compared to the one applied for the heat exchanger and the ceramic TWC (1.5 $W/m^2/K$) since the channel is not assumed to be covered by the interior insulation. The heat transfer coefficient between exhaust gas and channel walls is valid for the range of mass flows occurring during the NEDC sequence

Table 4.3.: Parameters for lumped model of side channel between heat exchanger and catalyst in sequential system.

Total wall surface A_{sc}	$[m^2]$	0.0867
External wall surface A_{sc}^{ext} (without bottom plate)	$[m^2]$	0.0624
Channel weight m_{sc}	$[kg]$	0.676
Heat loss coeff. α_{sc}^{amb}	$[W/m^2/K]$	5.0
Heat transfer coeff. gas/wall $\alpha_{sc}^{g,s}$	$[W/m^2/K]$	10

assuming constant temperature (500°C). It turns out, that under these conditions the flow through the side channel is laminar resulting in an average heat transfer coefficient well below the one obtained in the narrow heat exchanger channels [27].

As for the integrated prototype presented in the previous section, a temperature-controlled electric heater is implemented which operates during the initial bypass phase and whose energy output is controlled based on a certain set point temperature. Yet, since the exhaust passes through the heater in both operating modes it can also operate when the bypass flap is closed. For instance, the cooling effect of the side channel, which occurs after transition to countercurrent operation, can be balanced by prolonged heating. This strategy is investigated in Section 4.3.2.3.

4.3.2.2. NEDC simulation results, equal flap strategy as for integrated system

In order to assess the benefits of the sequential design approach, NEDC simulations are conducted and the results are compared to those obtained with the integrated system. Equal length of the initial heating period as previously determined for the integrated system is assumed while two different temperature set points are tested. The resulting transient evolution of average catalyst temperature, as well as accumulated CH_4 emissions and heating power, are shown in Figure 4.14. During the initial heating phase a clear advantage of the sequential system emerges due to the lower thermal mass of the ceramic monolith brick and the higher specific surface area available for catalytic reaction. Consequently, full conversion of CH_4 can be achieved much sooner leading to significantly decreased tailpipe emissions. As soon as the bypass flap is closed and the heater switched off, the catalyst temperature decreases in case of the sequential system due to the above described cooling effect of the side channel. However, if the same heating strategy as for the integrated system is applied, the temperature level obviously remains above light-off throughout the remaining cycle resulting in very low accumulated emissions. The controller set point of the electric heater can therefore be reduced from 750°C to 450°C, resulting

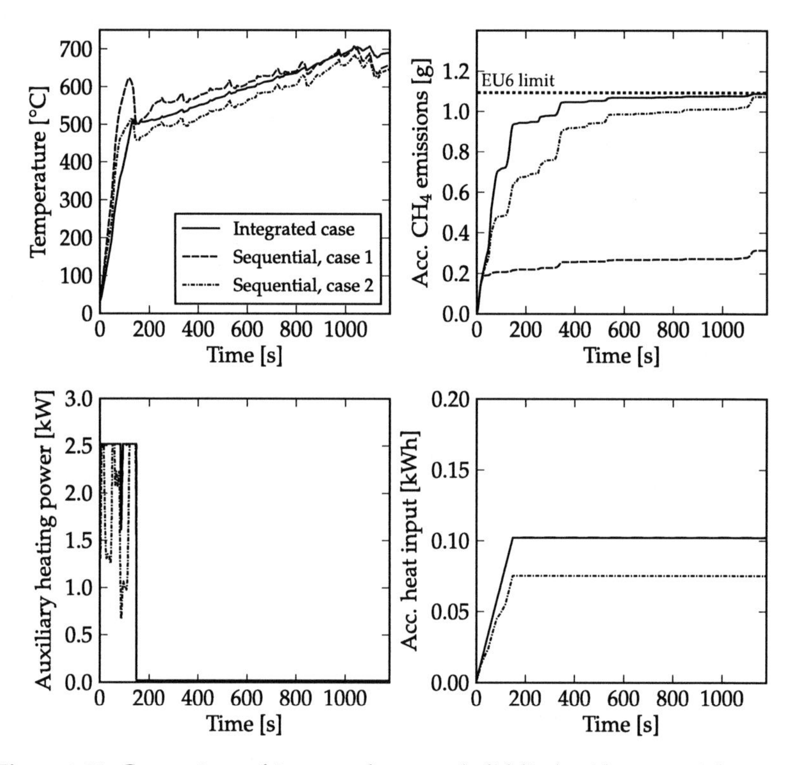

Figure 4.14.: Comparison of integrated system (solid line) with sequential system and two different heating strategies. Case 1 (dashed line): equal heating period and set point temperature ($t_s = 147\,s$, $T_{set} = 750°C$) as integrated case. Case 2 (dash-dotted line): decreased setpoint temperature ($t_s = 147\,s$, $T_{set} = 450°C$).

in a decreased power input during the heating phase (case 2). Then, the catalyst temperature is still close to light-off when the flap is closed. For this reason, further CH_4 emissions occur until the catalyst temperature exceeds 500°C. Eventually, a very similar emission level as in case of the integrated system is obtained. Yet, the required heating energy is about 25% lower. Further optimization of the conversion performance focusses on avoiding the temperature drop after transition to countercurrent operation in order to maintain the high temperature level obtained during the preceding heating phase. The resulting strategy is presented in the following section.

4.3.2.3. Optimization of Flap operating strategy

In Figure 4.14 it is shown that a sharp transition from heating mode to countercurrent operation leads to a drop of average catalyst temperature due to the cold side channel. As compensation, a bypass flap with flexible closing speed is assumed in the following. The closing speed is characterized by a scaling factor s_F, resulting in a modified formulation of the previously shown equation of an ideal binary flap (Equation 4.1):

$$F(t) = 0.5 \cdot (1 - tanh\,(s_F \cdot (t - t_s)))\,, \tag{4.4}$$

with $s_F \leq 1$. In addition to that, the heater is assumed to remain active as long as the bypass ratio is greater than zero. For the simulation results shown in the following, the same heating strategy as the one applied for "case 2" in the previous section (i.e. $t_s = 147\,s$, $T_{set} = 450°C$) is used. In Figure 4.15, the results obtained with three different values of s_F are compared to the original approach (dash-dotted line in Figure 4.14 where $s_F = 1$). Obviously, the magnitude of the temperature drop can be successfully decreased by closing the flap more slowly and by slightly extending the heating support. With $s_F = 0.1$, methane tailpipe emissions can be reduced by $\sim$ 20% while the required energy input only increases by 12%. For even lower values of s_F the gains in terms of tailpipe emissions are rather small. This especially applies for $s_F = 0.01$, representing the lowest flap closing speed evaluated, where the increase of energy input has about the same magnitude as the achieved decrease of emissions.

4.4. Conclusions

In order to quickly heat the converter's catalytic zone to the required ignition temperature, the flap/bypass system proved to be clearly beneficial. With the coated zone above CH_4 light-off temperature and the system operated in countercurrent mode, the catalyst temperature can be maintained or even increased during the sim-

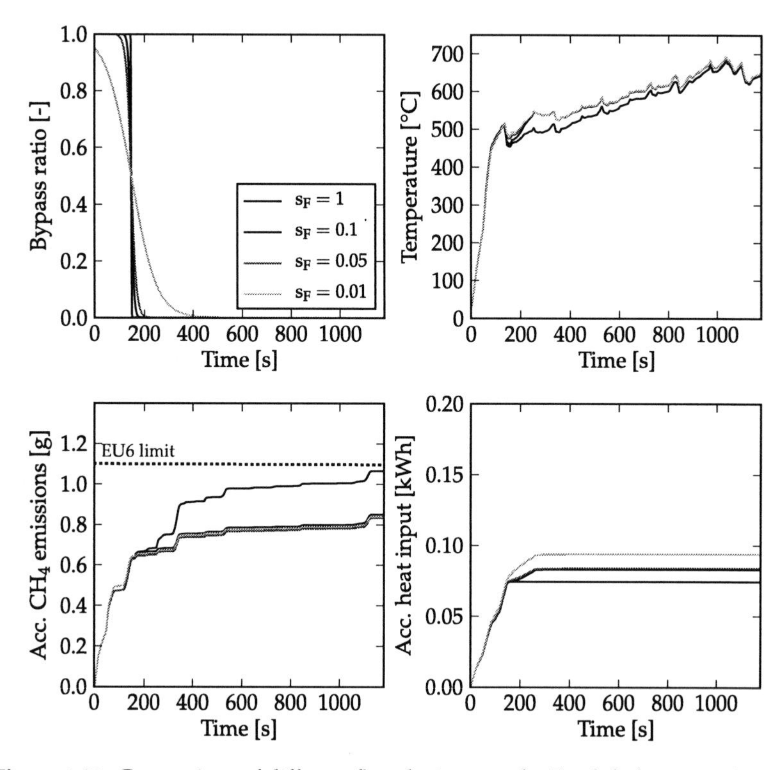

Figure 4.15.: Comparison of different flap closing speeds. Top left: bypass ratio versus time. Top right: mean catalyst temperature for different flap strategies. Bottom left: accumulated CH_4 emissions. Bottom right: accumulated energy input.

ulated NEDC sequence. Moreover, in case of high exhaust temperature the flap can be re-opened to deactivate the heat exchanger's amplification effect and protect the system from over-temperatures. Regarding the bypass flap's operating strategy, a binary approach with a specific switch time, where the flap position changes from fully opened to fully closed, was found to be sufficient. It turned out, that for lowest emission results the complete catalyst zone should be heated above the ignition temperature of CH_4 before the flap closes. Further gains in conversion efficiency are possible by a flexible bypass operating strategy. In this case the flap re-opens during acceleration phases within the first 400 seconds of the cycle when the exhaust temperature exceeds the level required for methane light-off.

Although the reference case, featuring a coated ceramic honeycomb monolith with equal cell density and catalyst volume as the heat exchanger's coated part, heats up much more quickly, it falls behind the performance of the heat-integrated system during the remaining drive cycle. This is due to a decrease of catalyst temperature during each idling phase. Hence, slip of CH_4 occurs when the vehicle accelerates afterwards. In order to achieve similar emission results as with the heat-integrated approach, increased precious metal loading or catalyst volume are required.

In combination with auxiliary power input by an electric heater the converter can be placed remote of the engine in an underfloor position where the exhaust temperature is significantly lower compared to a close-coupled position. Due to the heat exchanger's amplification effect, additional heating input is only required during the initial phase of the drive cycle.

The simulation results showed, that a major drawback of the heat-integrated system is the high thermal mass of the metallic catalyst substrate. As a result, light-off is delayed leading to higher THC emissions during initial catalyst heating. Moreover, only half of the coated volume (i.e. outflow channels of the heat exchanger) is passed by the exhaust during heating mode. Therefore, the largest portion of THC emissions occurs during the heating phase of the drive cycle which is why further development concentrated on optimizing the system's performance during the initial phase of the drive cycle.

As most promising option the sequential concept, comprising a standard ceramic honeycomb with catalytic coating and an inert metallic heat exchanger, emerged. Due to the ceramic brick's low thermal mass, CH_4 light-off occurs significantly faster than in case of the fully metallic integrated system. In fact, heating occurs equally fast as in case of the standard ceramic monolith. In addition to that, much more flexible combinations of catalyst substrates with high cell density and heat exchangers with backpressure-optimzed layout are feasible. However it was shown, that the exterior side channel between heat exchanger and catalyst leads to some decrease of performance during transition from heating mode to countercurrent operation. Besides the possibility to integrate this channel into the casing's insulation jacket, the flap operating strategy can be optimized to preheat the side channel during the initial heating phase.

Chapter 5.

Reactor prototypes and experimental evaluation

In this chapter, the laboratory-scale prototypes developed and built for experimental testing are presented. In each case, specific experimental results are shown and discussed in order to illustrate the development of the heat-integrated concept towards a system suitable for dynamic operation in a vehicle.

Initially, a catalyst coated, metallic heat exchanger based on the previous folded-sheet design approach [5, 24, 25] is presented and the results obtained during stationary testing under fuel-lean conditions are discussed. The performance of this system is regarded as benchmark for the newly developed brazed heat exchanger which is presented subsequently. This prototype aimed at achieving a layout which is based on industrial manufacturing standards and suits large-scale production. At the same time, its performance should reach the values achieved with the benchmark prototype. Therefore, the experimental results obtained during fuel-lean stationary testing are compared to the previous folded-sheet heat exchanger. In addition, stationary tests were conducted with stoichiometric gas composition in order to evaluate the influence of air-fuel ratio on methane conversion.

Although the brazed heat exchanger is expected to well perform under stationary conditions, its high thermal mass hinders rapid catalyst heating during cold start. In the preceding chapter, it is shown by drive cycle simulations that this leads to an increased heating demand, especially if the system is mounted in an underfloor position (see Section 4.2.4). Therefore, a sequential heat-integrated system was developed which comprises a standard ceramic honeycomb catalyst with high cell density and low thermal mass and a metallic countercurrent heat exchanger (see Section 4.3). According to drive cycle simulations conducted with this setup (see Section 4.3.2.2), cold start emissions can be significantly decreased if the same heating power input as for an integrated heat exchanger system with equal global dimensions is assumed. While there is a clear benefit during catalyst heating in bypass mode, the transition to countercurrent operation remains challenging with this setup. The simulation results reveal that when the bypass flap closes, a pronounced drop of catalyst temperature occurs which is due to a cooling effect of the lateral connection channel between heat exchanger and catalyst. In this chapter, results of transient cold start experiments with a laboratory-scale sequential system are shown which confirm these findings.

5.1. Folded sheet prototype

5.1.1. Reactor layout and dimensions

For the initial set of experiments, a metallic heat-exchanger prototype according to the previously developed folded-sheet design approach [5, 24, 25] was applied. Alternative reactor configurations following the same heat exchanger layout were published elsewhere [31]. Figure 5.1 shows a photograph of such a heat exchanger in a transparent casing in order to illustrate its functional principle.

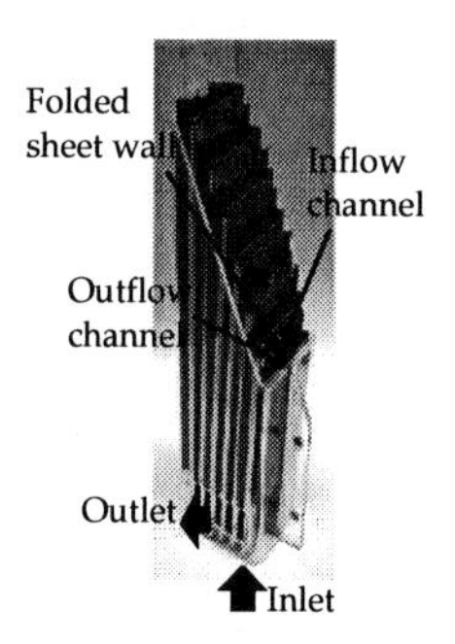

Figure 5.1.: Model setup of folded-sheet heat exchanger according to [24].

Inflow and outflow channels are separated by a meandering folded metal sheet. Corrugated spacer structures are placed between each fold for improved mechanical stability and enhanced thermal contact between gas and wall. Furthermore, these spacers provide increased geometric surface for catalyst coating. To insulate the open lateral faces of the folded metal sheet from each other, the structure is wrapped by a ceramic fiber sheet, which is tightly compressed by the outer casing. While the inflow channels at the lower end remain open, the outflow channels have to be sealed and a sidewise outlet has to be provided. Since very thin metal sheets are used for the folded sheet wall in order to minimize thermal mass, the heat input during fusing of the channel ends has to be controlled precisely which can be realized by laser welding.

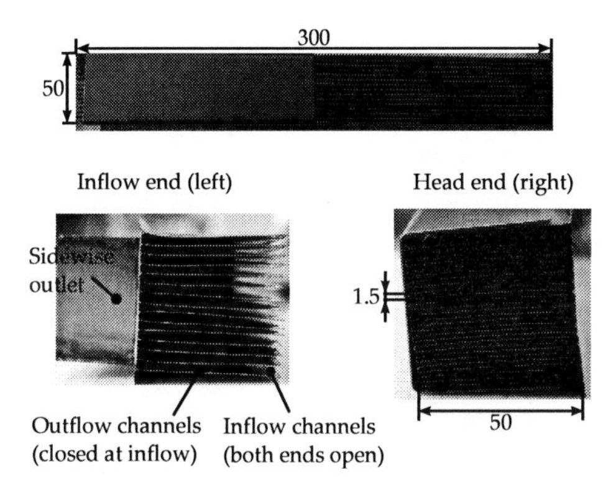

Figure 5.2.: Heat exchanger core of folded-sheet prototype prior to coating and insulation / canning with dimensions. Top: side view. Bottom left: inflow end with closed outflow channels, sidewise outlet port and open inflow channels (spacers omitted). Bottom right: opposite end of heat exchanger where gas passes from inflow to outflow channels.

In Figure 5.2, photographs of the final laboratory-scale prototype prior to coating and canning are depicted with according geometric dimensions. For improved sealing of the left-hand side lateral face, a thin metal sheet with rectangular port attached at the sidewise outlet was laser welded to the seams of the folded sheet structure. In the bottom left picture of Figure 5.2, the inflow end with closed outflow channels is depicted. The sidewise outlet, which is not visible from this perspective, was achieved by spacer slices with tilted corrugation (see Figure 1.1, bottom). As can be seen in this picture, the tight welding of the thin wall material and crotches between folded sheet and side plate is delicate. This renders the folded sheet setup rather inappropriate for mass customization on a large industrial scale.

In Table 5.1, the geometric properties of the folded-sheet prototype are enlisted. Stainless steel 1.4767 was used for both the wall and the corrugated spacers. The heat exchanger core was coated with a commercial Pd/Rh (39:1) three-way catalyst [9] which is optimized for exhaust gas aftertreatment of CNG powered vehicles. The coating zone stretches over 100 mm measured from the right end of the heat exchanger. After coating, 10 thermocouples of type K (NiCr-Ni) with 0.5 mm of outer diameter were distributed along the central axis of the heat exchanger. In Figure 5.3, a picture of the final reactor prototype with insulated casing and flanges for attachment to the test rig is depicted.

Table 5.1.: Geometric properties of folded-sheet heat exchanger.

Total length	$[m]$	0.3
Coated length	$[m]$	0.1
Cross-sectional area	$[m^2]$	0.0025
Channel height	$[mm]$	1.5
Sheet thickness wall	$[\mu m]$	110
Sheet thickness spacer	$[\mu m]$	50
Cell density	$[cpsi]$	385

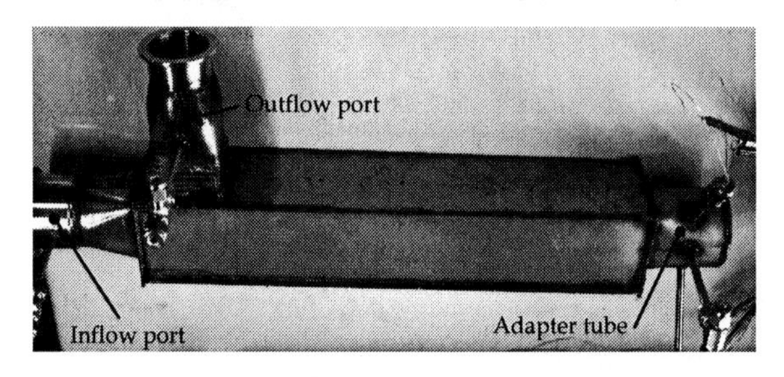

Figure 5.3.: Final lab-scale, folded-sheet prototype fitted with thermocouples and insulated casing. The adapter tube (right) was attached for optional retrofitting of a cold start heater.

An adapter tube was attached to the right end of the heat exchanger for retrofitting of a cold start heater. However, for the stationary experiments shown in the following, the tube was closed and, as the complete heat exchanger, covered with additional external insulation. Further details on the experimental facilities used for stationary testing are shown in Appendix A.

5.1.2. Stationary experimental evaluation

As base flow for the stationary measurements, heated pressurized air was used with traces of methane added as fuel. While the inflow temperature was kept constant during each experimental run, the base flow could be varied between 4 and 25 Nm^3/h, corresponding to $GHSV$ between 5000 and 33000 h^{-1} with the total heat exchanger volume as reference. The first set of results shown in the following section was obtained with constant feed concentration of CH_4 (i.e. constant adiabatic temperature rise), which, according to Equation 2.38, leads to different maximum temperatures during variation of $GHSV$. Due to the relatively long stabilization time required for obtaining sufficiently stationary results at each value of volumetric flow, an automated methane dosage system was set up for the subsequently conducted experiments [64]. It turned out that if T_{in} and T_{max} are kept constant, the axial temperature profile undergoes only small changes in the partially coated system when the flow rate is altered. This tremendously reduces the required stabilization time. Moreover, the system is protected from drifting into regions of excessive maximum temperature which might occur especially under low throughput conditions (see Section 3.2.3). T_{max} is kept constant by controlling the methane feed concentration.

5.1.2.1. Results without heat loss compensation at adapter tube

In order to start catalytic conversion of methane on the catalyst, the system was heated by the hot base flow ($GHSV = 12500\,h^{-1}$) starting from ambient temperature. Additionally, 6700 ppm of hydrogen ($\Delta T_{ad} \approx 50$ K) were added and the feed temperature was set to 300°C. Due to the heat liberated by hydrogen conversion and amplification by the heat exchanger, the catalyst could be heated above the required ignition temperature of methane (see Section 3.2.1). Then, hydrogen was shut off and dosage of CH_4 was activated until the stationary state was reached. For practical reasons, stationary conditions were assumed when the temporal gradient of all axial temperatures was below 0.1 K/min. Then, the measurements were recorded and the base flow rate was adjusted to the next value. In Figure 5.4, axial profiles of concentration and temperature are shown for three different space velocities. For these experiments the inflow concentration of methane was fixed at 3500 ppm (i.e. no automatic adaption of methane feed). In each of the shown cases, the experimental values (markers in Figure 5.4) are plotted together with profiles calculated with the fitted 1D simulation model. Lower maximum temperature can be observed for higher flow rates indicating a decrease of amplification factor. Total methane conversion is almost completely independent of the respective space velocity. Compared to the ideal result in a partially coated countercurrent reactor (see Figure 2.7), the shape of the axial temperature profile is rather peculiar especially at the lowest flow rate (Figure 5.4, top row). At this operating point, a pronounced decrease of temperature over the coated zone can be observed. This phenomenon arises from the exposed adapter tube which obviously acts as strong heat sink despite its external insulation. In order to adapt the simulation model to the experimental values, the following mathematical expression mimicking the influence of this heat sink was implemented at the transition between inflow and outflow channels:

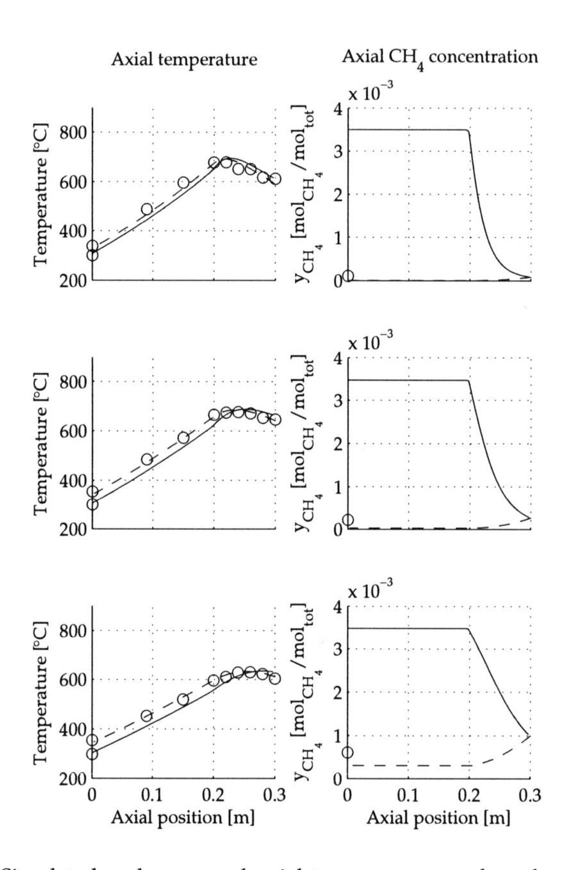

Figure 5.4.: Simulated and measured axial temperatures and methane concentrations for $GHSV = 9000\ h^{-1}$ (top row), $GHSV = 15000\ h^{-1}$ (center row) and $GHSV = 20000\ h^{-1}$ (bottom row). Solid lines: calculated profiles in inflow channels. Dashed lines: calculated profiles in outflow channels. Markers: measured values (folded-sheet prototype without heat loss compensation).

$$T^{in,outflow} = T^{out,inflow} - \frac{\alpha^{amb}_{tube} A}{\dot{m}^g \bar{c}_p} \left(T^{in,outflow} - T^{amb} \right). \tag{5.1}$$

Equation 5.1 models the tube as ideally mixed stationary CSTR with the enthalpy fluxes of in- and outflowing exhaust and heat loss over the walls. All parameters except of the heat transfer coefficient α^{amb}_{tube} are known or can be calculated based on the chamber's simple tubular geometry. However, the heat transfer coefficient is difficult to determine since the flow conditions inside the chamber are rather complex and also likely to change over the investigated range of flow rates. Hence, this parameter was fitted during model adaptation. As additional fitting parameters, the global heat loss over the heat exchanger casing and the pre-exponential factor of the methane oxidation rate were used. The activation energy was not optimized since the catalyst temperature was generally above the light-off range during stationary testing. In order to efficiently determine the optimal parameter set, the DIANA simulation model was embedded into an optimization script using the COBYLA (Constrained Optimization By Linear Approximation) algorithm [61]. This solver is part of the SciPy package [32]. The final parameters are shown in Table 5.2.

Table 5.2.: Fitted parameters for folded-sheet prototype without heat loss compensation at adapter tube.

$\alpha^{s,amb}$ $\left[\frac{W}{m^2K}\right]$	α^{amb}_{tube} $\left[\frac{W}{m^2K}\right]$	$k^0_{CH_4}$ $\left[\frac{kmol}{m^2sbar}\right]$
1.55	16.0-24.0	6.0

Compared to the original parameter set (see Section 2.1.1.7 and Appendix D), the global heat loss coefficient was increased by 3%. The pre-exponential factor was decreased considerably which might indicate relatively low catalyst activity. Yet, it is not clear whether this is due to the poor mechanical stability of the washcoat on the metal surface or if the catalyst surface was reduced due to partially blocked channels arising from the dip-coating process. The fitted heat transfer coefficient of the adapter tube α^{amb}_{tube} was found to vary over the observed range of flow rates. As discussed above, this can be explained by varying flow conditions inside the chamber which are not captured by the lumped modeling approach. In any case, the above mentioned results underline the significant impact of heat losses in the hot part of the heat exchanger on the system's stationary performance.

5.1.2.2. Results with heat loss compensation at adapter tube

In a second experimental run, the adapter tube was wrapped with a temperature-controlled heating wire. In this case, the methane feed concentration was controlled in order to keep the maximum temperature constant at 630°C which was also the set-point for the heating wire. As for the initial case without compensation, the model was fitted to the experimental results with the heat transfer coefficient at adapter tube α_{tube}^{amb} and the pre-exponential factor of the methane conversion rate as fitting parameters. The latter parameter was again optimized since the catalyst was expected to have undergone some thermal aging due to operation above 600°C during the preceding stationary experiments. In order to prove that the heat losses are not over-compensated by electric heating, calculations with adiabatic adapter tube (i.e. $\alpha_{tube}^{amb} = 0$ in the simulation model) were conducted. In Table 5.3, the optimized parameters are shown while the results are depicted in Figure 5.5.

Table 5.3.: Fitted parameters for folded-sheet prototype with heat loss compensation at adapter tube.

$\alpha^{s,amb}$ $\left[\frac{W}{m^2K}\right]$	α_{tube}^{amb} $\left[\frac{W}{m^2K}\right]$	$k_{CH_4}^0$ $\left[\frac{kmol}{m^2sbar}\right]$
1.55	2.0-6.0	4.0

Obviously, the shape of the axial temperature profiles approaches the ideal case without exceeding the maximum temperatures calculated with adiabatic adapter tube. The values of α_{tube}^{amb} range well below the ones obtained during model regression of the experiments without heat loss compensation. However, the pre-exponential factor also decreased considerably indicating relatively strong catalyst deactivation. At this point of the study the main reason of this phenomenon was not yet clear. A more detailed discussion will be given in Section 5.2.4 after having also shown the stationary performance of the brazed heat exchanger prototype.

Subsequently, the stationary performance of the folded-sheet type heat exchanger was evaluated by calculating the characteristic numbers introduced in Section 2.2. In order to obtain the amplification factor a_F, the maximum temperature rise ΔT_{max} has to be divided by the respective value of ΔT_{ad} which is determined by the amount of methane converted on the catalytic coating. In Figure 5.6, the resulting values of both experimental cases (i.e. with and without heat loss compensation at the adapter tube) are plotted over the complete range of observed space velocities and compared to the theoretic result according to an evaluation of Equation 2.38 with appropriate geometric parameters. Obviously, heat losses cause a considerable deterioration of heat exchanger performance throughout the complete range of evaluated space velocities. However, the characteristic shape of the curve is confirmed by the experimental results. At low space velocities, a_F increases until reaching the maximum where the influence of axial heat conduction in the solid and heat transfer is just balanced. In fact, at this point the sum of Pe^{-1} and NTU^{-1} is minimal. At higher space velocities, Pe increases and NTU decreases leading to lower values of a_F. The values

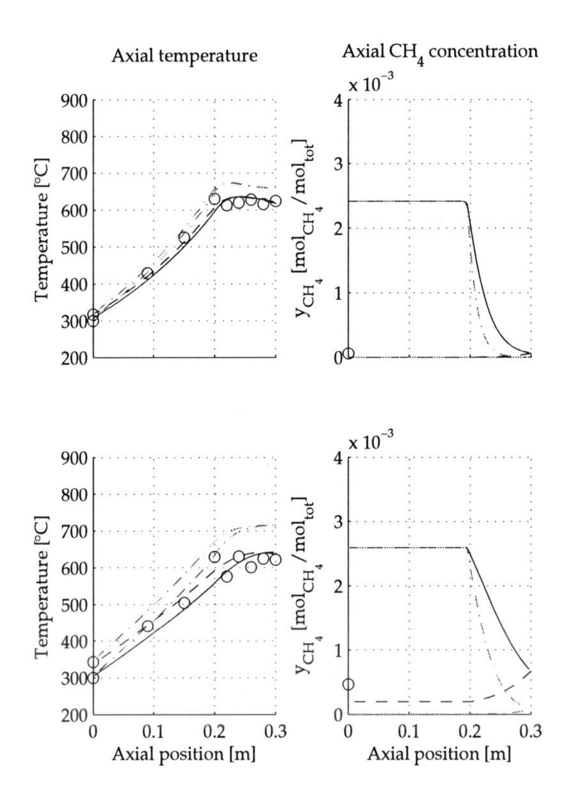

Figure 5.5.: Simulated and measured axial temperatures and methane concentrations for $GHSV = 5000\,h^{-1}$ (top row) and $GHSV = 20000\,h^{-1}$ (bottom row), folded-sheet prototype with heat loss compensation. Solid black lines: calculated profiles in inflow channels. Dashed black lines: calculated profiles in outflow channels. Markers: measured values. The gray dash-dotted lines represent the simulation case with adiabatic adapter tube.

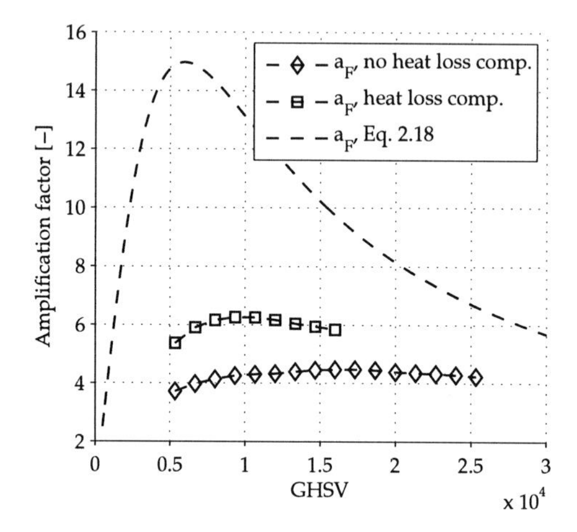

Figure 5.6.: Comparison of experimentally obtained amplification factors of the folded-sheet prototype and theoretical result excluding external heat losses.

obtained in this study are below previously published results with similar heat exchanger prototypes [5]. However, in that case the coated length was shorter at equal total heat exchanger length (8 cm versus 10 cm in the present case) and the inflow temperature was considerably lower. Further experiments with decreased inflow temperature (50°C) revealed, that maximum amplification factors of 8 are possible with the folded sheet prototype which is the same order of magnitude as in case of the results published in [5].

The second, more common performance parameter is the heat exchanger efficiency η. By Equation 2.34, it can be directly related to the amplification factor assuming ideal conditions. Due to the influence of radiation by the channel walls on the temperature sensors in the narrow heat exchanger channels, calculation of η based on a relation of gas phase temperatures is expected to be rather imprecise. In Table 5.4, a compilation of measured amplification factors and efficiencies calculated with Equation 2.34 is given for the case with heat loss compensation.

Table 5.4.: Amplification factors a_F and heat exchanger efficiencies η in the observed range of space velocities for the folded-sheet heat exchanger type with heat loss compensation at the adapter tube.

$GHSV \cdot 10^3 \left[\frac{1}{h}\right]$	a_F [−]	η [%]
5	5.4	81
11	6.2	84
15	5.8	83

5.2. Brazed prototype

While the stationary performance obtained with the folded-sheet prototype was promising, some drawbacks of this design approach emerged. First of all, the relatively flexible structure leads to problems during dip-coating where the catalyst is applied in multiple steps. Between each process stage the substrate is heated for drying and calcination of the washcoat. This leads to thermal strain in the washcoat since at this point the heat exchanger package is not yet compressed and fixed in the final casing. As a result, adhesion of the catalyst coating on the metallic wall is relatively poor, thus leading to some chipping of washcoat during subsequent operation of the heat exchanger. Furthermore, the manufacturing process of the folded sheet structure is relatively complex, particularly with regard to production on a large industrial scale.

5.2.1. Reactor layout and dimensions

For these reasons, a vacuum-brazed heat exchanger design was developed in collaboration with industrial partners aiming at achieving similar performance as with the folded-sheet heat exchanger prototype. At the same time, the manufacturing process should be simplified significantly. In Figure 5.7, photographs of the resulting laboratory-scale system with according dimensions are depicted.

The inflow channels are formed by flat tubes which are open at both ends. The corrugated spacer structures placed within are tightly brazed to the surrounding walls which improves both mechanical stability and thermal contact. At reactor inlet a header plate with openings for the inflow channels is attached which covers the outflow channels. This plate also serves as rim for the later welding of the heat exchanger core to the casing. The outflow channels are formed by corrugated spacer structures which are placed between the inflow channels and brazed to the neighboring walls. The sidewise outlet is achieved by short spacer slices with tilted corrugation (see Figure 1.1). Compared to the folded-sheet prototype the channel height is significantly increased which is due to the industrial partner's constraints in the

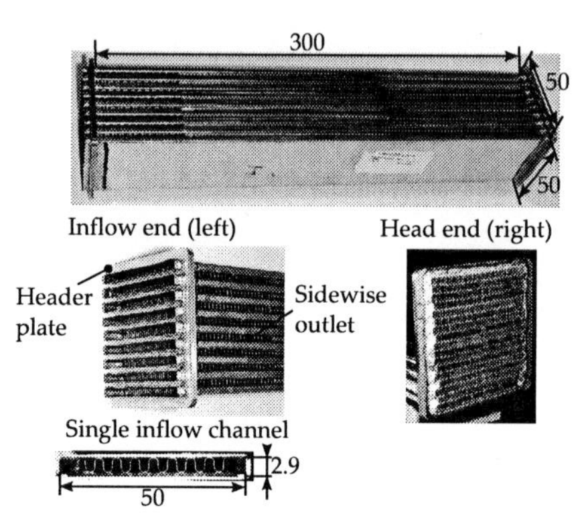

Figure 5.7.: Heat exchanger core of brazed prototype prior to coating and insulation / canning with dimensions. Top: side view. Bottom left: inflow end with header plate separating inflow and outflow channels and sidewise outlet. Bottom right: opposite end of heat exchanger where gas passes from inflow to outflow channels.

bending process of the flat tube's wall plates. This leads to relatively high hydraulic diameter and low cell density. While this is not critical in the inert part of the heat exchanger, the resulting geometric surface is rather low for catalyst coating. Hence, the cell density was increased in the active part of the heat exchanger by applying spacer structures with decreased fin pitch. The difference can be seen by looking at the inflow- and head end in Figure 5.7. As material for both the channel walls and the spacer structures stainless steel 1.4435 was applied.

In Table 5.5 the geometric properties of the laboratory-scale brazed prototype are listed.

Due to the lower cell density in the coated part, the coated zone was slightly prolonged compared to the folded-sheet prototype. This should however have no negative impact on the heat exchanger's amplification performance since the reaction zone shifts axially depending on the respective feed temperature (see Chapter 3). During catalyst coating, the high mechanical stability of the heat exchanger core led to significantly improved adhesion between washcoat and metallic substrate. The brazed channel stack was fitted with an insulated casing and an adapter tube for optional retrofitting of a cold start heater was attached to the right end. Contrary to the folded-sheet prototype, the tube was equipped with an internal insulation jacket for minimized heat loss. Finally, 10 thermocouples of type K (NiCr-Ni) with 0.5 mm of outer diameter were placed along the central axis of the heat exchanger.

Table 5.5.: Geometric properties of lab-scale brazed heat exchanger.

Total length	$[m]$	0.3
Coated length	$[m]$	0.12
Cross-sectional area	$[m^2]$	0.0025
Channel height	$[mm]$	2.9
Sheet thickness wall	$[\mu m]$	150
Sheet thickness spacer	$[\mu m]$	150
Cell density uncoated / coated	$[cpsi]$	133 / 250

5.2.2. Stationary experimental evaluation

The prototype was tested at the same test rig as previously used for stationary evaluation of the benchmark prototype (see Section A.1.1 for further details). Since the insulation of the adapter tube was significantly improved compared to the folded-sheet prototype, compensation heating was omitted. For reactor heat-up, the same strategy based on conversion of hydrogen until reaching CH_4 light-off was applied. During the subsequent variation of base flow rate at constant feed temperature (300°C), the maximum temperature was kept at 630°C with the methane feed concentration automatically adjusted. The resulting axial profiles of temperature, methane concentration and axial backpressure are depicted in Figure 5.8. The slightly kinked shape of the temperature and backpressure profiles arises from different cell densities in the uncoated and coated part of the heat exchanger. As a result, the geometric parameters (i.e. specific surface a_v, void fraction ϵ and hydraulic diameter) differ in the two parts which is reflected by a change of gradient in the axial profiles. In order to adapt the simulation model, the same technique with a heat sink at the hot end of the heat exchanger was assumed. The final set of adapted parameters is shown in Table 5.6.

Table 5.6.: Fitted parameters for brazed prototype

$\alpha^{s,amb}\left[\frac{W}{m^2K}\right]$	$\alpha^{amb}_{tube}\left[\frac{W}{m^2K}\right]$	$k^0_{CH_4}\left[\frac{kmol}{m^2sbar}\right]$	$\zeta_1\,[-]$	$\zeta_2\,[-]$
1.55	15.0	14.0	85.0	71.0

The resulting pre-exponential factor is significantly higher than the one obtained for the folded-sheet prototype. This is not surprising since the mechanical stability of the washcoat was obviously improved by the tightly brazed heat exchanger struc-

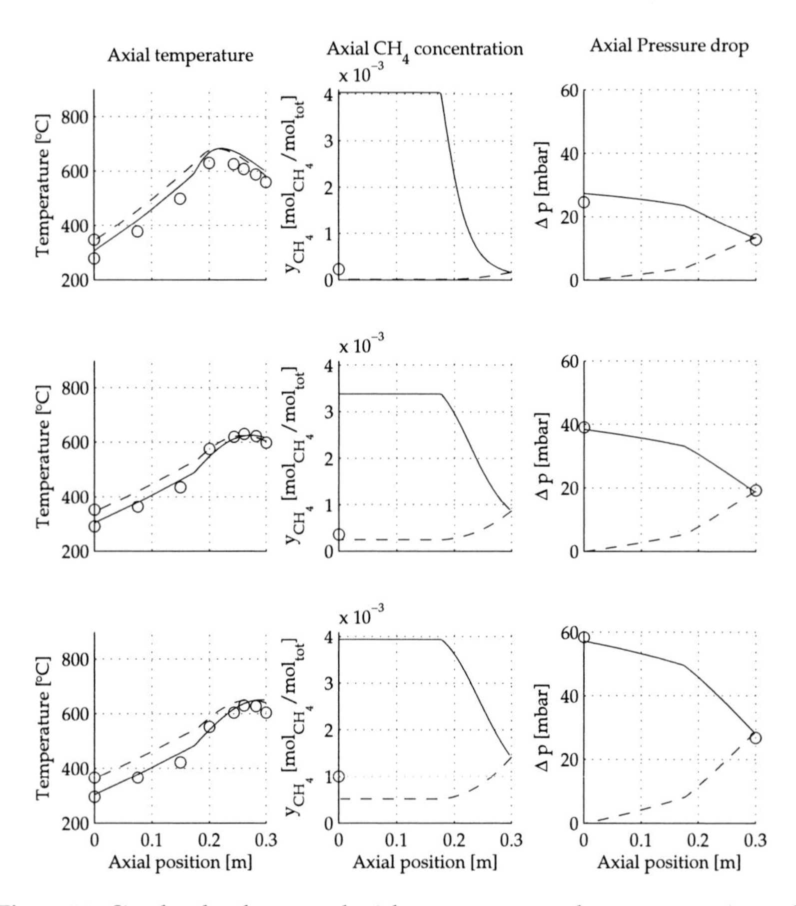

Figure 5.8.: Simulated and measured axial temperatures, methane concentration and axial backpressure for $GHSV = 9000\ h^{-1}$ (top row), $GHSV = 15000\ h^{-1}$ (center row) and $GHSV = 21000\ h^{-1}$ (bottom row). Solid lines: calculated profiles in inflow channels. Dashed lines: calculated profiles in outflow channels. Markers: measured values (brazed prototype).

ture. Although backpressure was modeled with rather simple expressions valid for laminar pipe flow (see Section 2.1.1.6), an adaption of the two drag coefficients ξ_1 and ξ_2 resulted in a decently accurate prediction of the measured Δp-values. Due to the improved insulation of the adapter tube, the heat loss parameter is significantly lower than the values previously obtained with the benchmark prototype (see Table 5.2). Moreover, the results depicted in Figure 5.8 were obtained with a constant value of α_{tube}^{amb}. Despite various changes in the geometric setup, the performance of the brazed heat exchanger seems to be very close to the above described benchmark represented by the folded-sheet prototype. A detailed comparison of both concepts will be therefore given in the following section.

5.2.3. Comparison of heat exchanger performance

In the first step, the methane conversion performance and axial backpressure of both prototypes was compared. The resulting plots are depicted in Figure 5.9. At low space velocities almost equal conversion is obtained with both prototypes. However, increased mass flows lead to a faster decline in case of the brazed prototype. Hence, as already discussed in Section 3.3.2, lower cell density in the coated part of the heat exchanger influences the conversion behavior especially under high-throughput conditions. As described for the folded sheet case, the amplification factor a_F can be calculated based on stationary measurements of the maximum temperature rise and the respective adiabatic temperature rise. In Figure 5.10 the results obtained with both prototypes for the case without heat loss compensation are shown. The brazed prototype is slightly inferior to the folded sheet one which is not surprising due to the lower overall cell density and the higher material thickness of both the spacers and the channel walls. The longer coated zone does not seem to have significant impact on the heat exchanger performance which is in accordance with the assumption of a self-adapting reaction zone described in Section 3.2.3. Around the maximum amplification factor very similar results were achieved with both prototypes. This finding underlines the beneficial effect of the tightly brazed heat exchanger core with excellent contact heat transfer between spacers and wall. Regarding backpressure, the brazed prototype is clearly advantageous due to the higher hydraulic diameter of the channels. In both cases, a linear dependence of backpressure on space velocities can be observed indicating laminar flow in the heat exchanger channels.

5.2.4. Stationary stoichiometric conditions

After having performed the stationary evaluation of both heat exchangers under fuel-lean conditions as described in the previous section, a very slow decay of methane

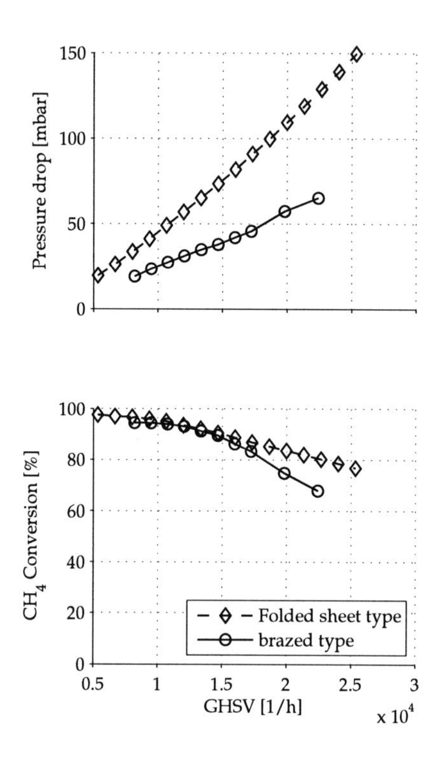

Figure 5.9.: Comparison of experimentally obtained backpressure (top) and methane conversion (bottom) for folded-sheet and brazed prototype.

conversion performance was observed with both prototypes. During evaluation of the folded-sheet heat exchanger, the poor coating quality was seen as main reason for decreasing catalyst activity. However, a similar effect also occurred during stationary evaluation of the brazed prototype with significantly improved washcoat stability. In order to quantify the deactivation, the brazed heat exchanger was repeatedly operated with equal space velocity ($GHSV = 10000\,h^{-1}$) and maximum temperature (630°C) after system start-up. The stationary conversion values over reactor operating time are depicted in Figure 5.11.

A very similar effect was observed during kinetic experiments with the same Pd/Rh-based catalyst material [8, 9]. The decay started as soon as the material was exposed

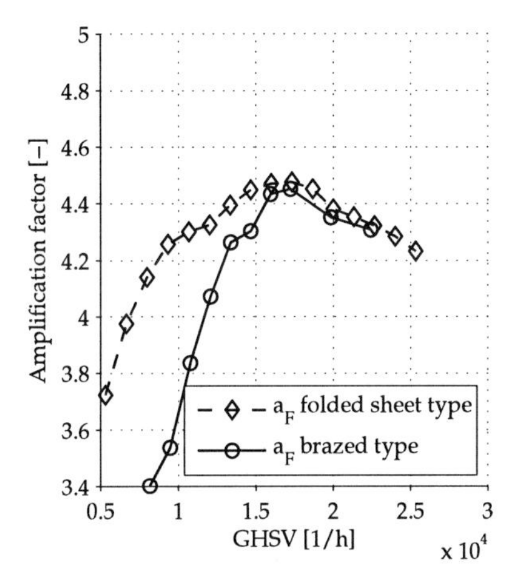

Figure 5.10.: Comparison of experimentally obtained amplification factor without heat loss compensation for folded-sheet (solid line) and brazed prototype (dashed line).

to stationary lean conditions (i.e. $\lambda > 1$) and could be reversed by oscillating gas composition around the stoichiometric point. In another study, it was shown that very short rich pulses distributed over relatively long lean phases were sufficient to regenerate the system [13]. A possible explanation for this behavior is that methane is believed to react with lattice oxygen, following a *Mars-van-Krevelen* mechanism [11, 33, 47]. For improved light-off performance, an optimal ratio between reduced (Pd) and oxidised (PdO_x) material is required to provide both the oxygen needed for methane oxidation as well as adsorption sites for the methane molecules [35]. The operation under nearly stoichiometric conditions obviously stabilizes the catalyst system whereas stationary lean operation leads to a slow transformation of the amorphous Pd and PdO_x species into PdO crystallites. The last mentioned species do not decompose as easily.

As a result, for further experiments with the heat-exchanger system a recuperative gas burner was used as exhaust gas generator. The basic features of this system are described in Section A.2 and a flowsheet of the final stoichiometric test rig is given in Section A.1.2. The slightly lean burner exhaust ($\lambda = 1.05$, $T_{out} = 180 - 300$°C), which did not contain any residual hydrocarbon species from the combustion of burner fuel, was mixed with a steady λ-controlled methane flow in order to achieve

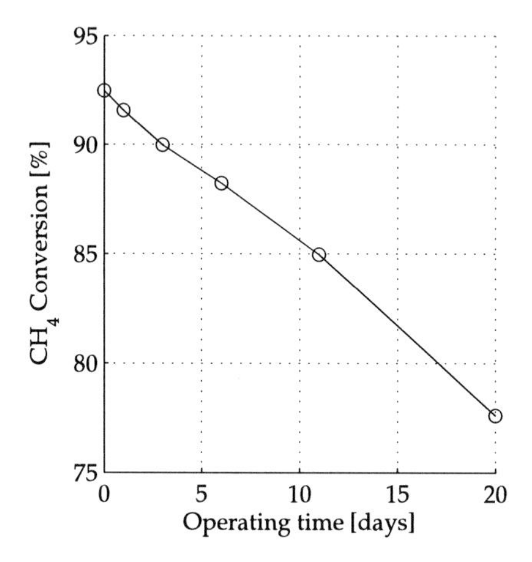

Figure 5.11.: Maximum conversion at constant maximum temperature (630°C) and space velocity (10000 h^{-1}) for the brazed prototype.

near-stoichiometric gas compositions before entering the heat exchanger. For reproduction of the above mentioned RedOx-phenomena, the brazed heat exchanger system was operated with slightly rich to stoichiometric exhaust gas compositions. For cold start and heating, hydrogen was used as described in the previous section until reaching sufficiently high catalyst temperatures. Since the methane concentration depends on the respective exhaust gas composition, the maximum temperature was not restricted in this case. In Figure 5.12, left column, results of methane conversion over space velocity are shown for slightly rich conditions ($\lambda = 0.99$). In addition, the respective methane concentrations at inflow and outflow, as well as the obtained maximum temperatures are shown. Obviously, for $\lambda = 0.99$ the initial activity of the catalyst is almost completely regained. At increased flow rates the methane feed concentration was slightly increased in order to keep the maximum temperature above 600°C. Since λ had to remain constant, the oxygen content in the burner exhaust was increased accordingly. Under these conditions with considerable amounts of water in the exhaust ($y_{H2O} > 10\%$), the activity of the catalyst appeared to be generally lower than with dry exhaust. This observation is supported by publications [12, 26] confirming the negative impact of water on CH_4 conversion efficiency of Pd-based catalysts.

In the next step, a λ-sweep was performed with constant burner exhaust flow rate and inflow temperature (Figure 5.12, right column). Due to the slow adaption of the axial temperature profile to any perturbation of gas composition, the waiting time at each point was about 30 minutes. The obtained dependence of methane conversion on λ was found to be very similar to previously published results with Pd-based catalysts [36]. Best methane conversion is obtained under slightly rich conditions while the transition to lean conditions almost immediately leads to a slow decay of activity. In the mentioned publication this deactivation effect was not observed since the time span covered by one sweep was considerably smaller. Close to the stoichiometric point the system reacts very sensitive to small perturbations of λ. This observation is also in accordance with the results discussed in [36].

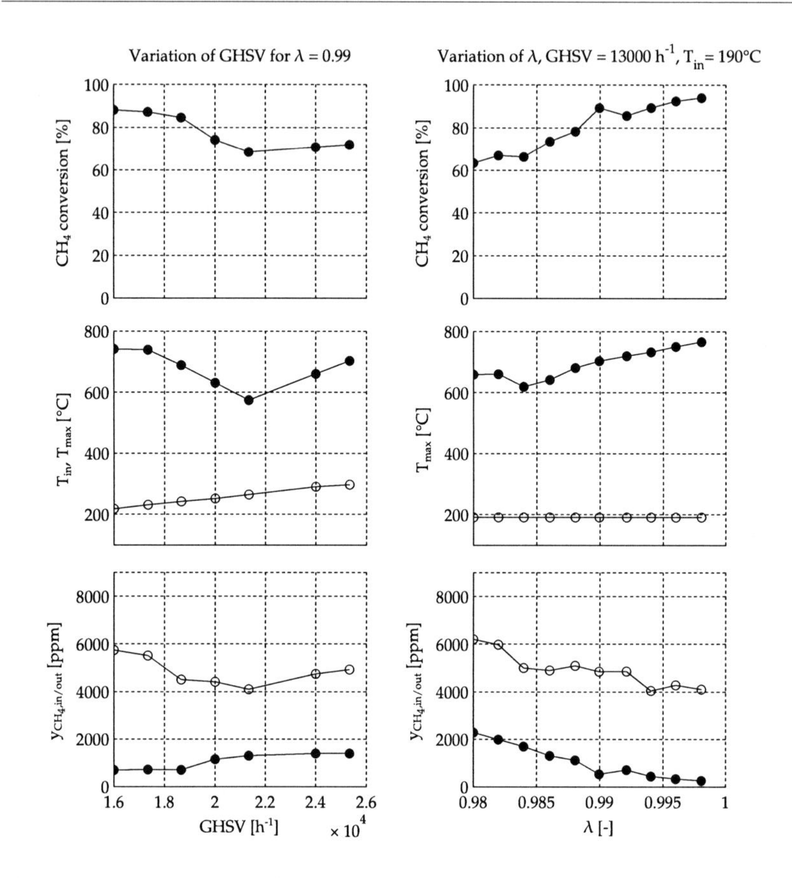

Figure 5.12.: Stationary results with near-stoichiometric exhaust gas composition. Variation of *GHSV* at constant λ (left column), Variation of λ at constant *GHSV* (right column). Open symbols: feed values, closed symbols: measurements at converter outlet (brazed prototype).

5.3. Sequential system - experimental evaluation

As shown in Section 4.2.3.2 by simulation results, the high thermal mass of the fully metallic, integrated system hinders heating of the catalytic zone during cold start. In the newly developed sequential system introduced in Section 4.3 however, the catalyst is coated on a standard ceramic honeycomb substrate with low thermal mass and combined with an inert metallic heat exchanger. As a result, the required light-off temperature for CH_4 conversion is reached much sooner. Furthermore, the sequential setup allows more flexibility regarding combinations of a catalyst brick with high cell density and a heat exchanger with wider channels for reduced backpressure. However, the simulation results also revealed a disadvantageous effect caused by the exterior side channel between heat exchanger and catalyst brick. During initial heating it is not passed by the exhaust and therefore remains relatively cold. Hence, as soon as the bypass flap closes and the system operates in countercurrent mode, the cold side channel walls act as heat sink on the passing exhaust gas. Possible countermeasures comprise prolonged heater operation and a flap with reduced closing speed (see Figure 4.15). In the following, the laboratory-scale prototype built up for experimental testing is presented together with specific experimental results in order to demonstrate the potential of this approach.

5.3.1. Reactor layout and dimensions

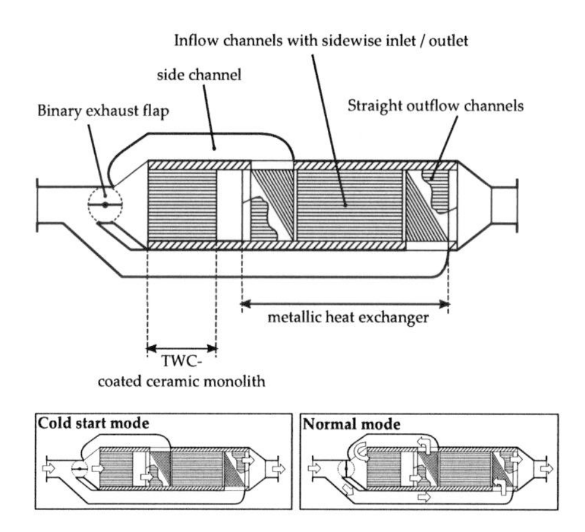

Figure 5.13.: Conceptual layout of sequential system for laboratory-scale experimental evaluation (top). Bottom left: flow configuration during cold start. Bottom right: normal mode flow configuration.

In Figure 5.13 the conceptual layout of the laboratory-scale sequential system is depicted with its main features highlighted. Contrary to the previous systems, the inflow channels of the heat exchanger have both sidewise inlets and outlets. The outflow channels are straight, which leads to decreased backpressure during bypass operation with open flap (i.e. straight flow through catalyst and outflow channels of heat exchanger). Contrary to the system layout assumed in Figure 4.3, the electric heater required for cold start could not be placed close to the catalyst brick since no system with rectangular cross section was available. Instead, a standard tubular EMICAT system [46, 59] was placed in an upstream position (see Figure 5.15 and Figure 5.16).

The metallic heat exchanger developed and built for the sequential system follows the basic design principles of the vacuum-brazed prototype presented in Section 5.2.1. Yet, while for that heat exchanger relatively high material thickness was required, thinner metal sheets were applied in the present case. Braze filler materials with lower melting point (700°C) and without the need of vacuum or inert gas atmosphere during processing were used. Certainly, the resulting heat exchanger is not as heat resistant as the vacuum-brazed one but this is not regarded critical for operation under well-defined laboratory conditions. In Figure 5.14, pictures of the resulting heat exchanger core are depicted. As for the previous brazed prototype,

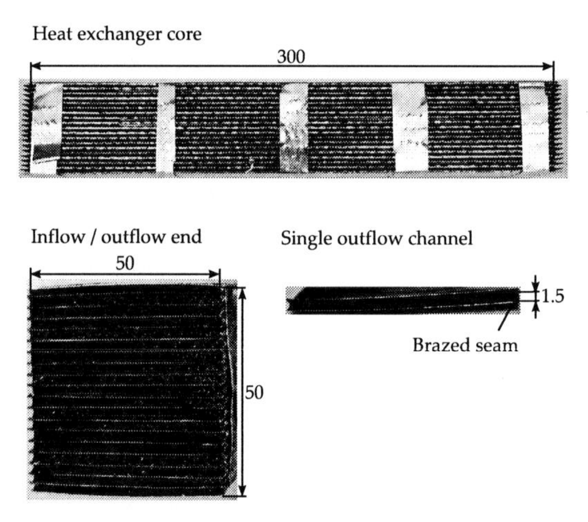

Figure 5.14.: Heat exchanger core for application in sequential system. Top: side view of stacked channels. Lower left: Frontal view. Inflow of axially open channels is straight while the closed ones are accessed over sidewise ports.

rectangular flat tubes are used for one flow direction. Corrugated spacers are placed inside for improved mechanical stability and thermal contact. Instead of using two wall plates with lateral brazing seams on either side as in case of the vacuum-brazed prototype, the present channels consist of a single metal sheet made of stainless steel 1.4767, which is folded lengthwise resulting in only one brazing seam (see Figure 5.14, lower right). Hence, less braze filler material is required resulting in lower thermal mass and a smaller risk of leakage due to brazing imperfections. Between the flat tubes corrugated spacers made of the same stainless steel are placed. Two sidewise ports with opposite orientation are realized by spacer slices with tilted corrugation (see Figure 5.13 and Figure 1.1, bottom), located at both ends of the heat exchanger. In order to close these channels axially, the flaps located at the ends of adjacent flat tubes are brazed together forming cuspid, aerodynamically efficient channel ends (Figure 1.1, bottom). After having completed the heat exchanger package, it was equipped with sealing belts at inflow and outflow end (see Figure 5.15 for further details).

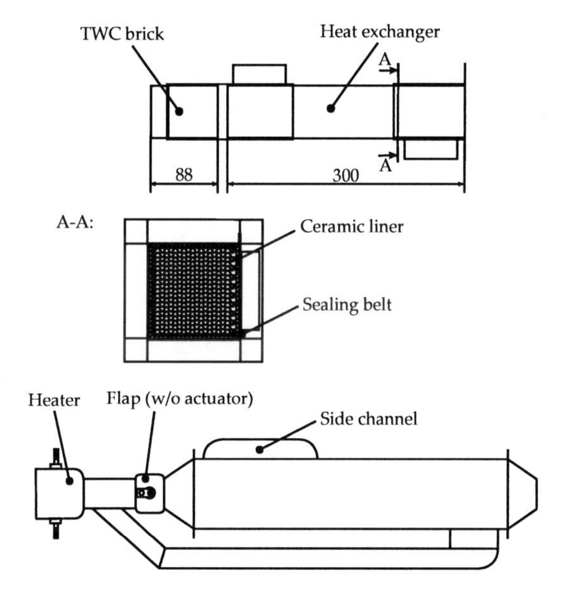

Figure 5.15.: Heat exchanger core with attached sealing belts (top). Sectional view (center) and complete assembly with flap and heater (bottom).

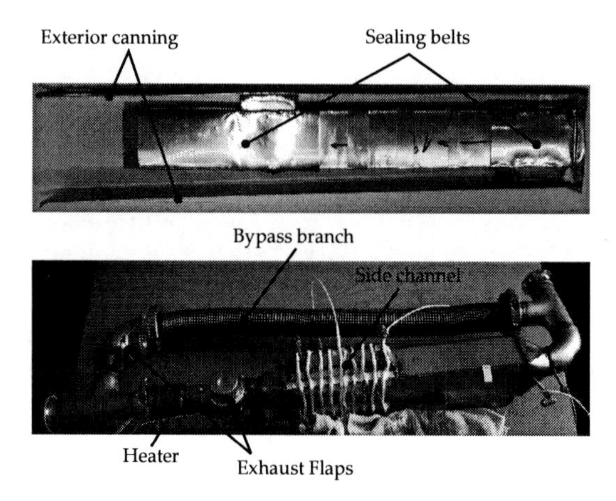

Figure 5.16.: Assembly of heat exchanger and catalyst prior to insulation and final canning (top). Bottom: canned prototype with heater and exhaust flaps.

The function of these belts is manifold: first of all they are used for tight compression of the heat-exchanger package and sealing of the sidewise outlets. The latter is achieved by a thin ceramic liner which is placed beneath the compression belts (Figure 5.15, center). Furthermore, rectangular frames attached to the belts are used as connection between heat-exchanger core and canning. In fact, a similar technique but with only one sidewise port was used for the initial folded sheet prototype (see Figure 5.1 and Figure 5.2). Finally, the left-hand side belt is used for fixation and sealing of the catalyst brick. In Figure 5.16, top, the final arrangement of heat exchanger and catalyst with attached sealing belts is shown. The dimensions of the laboratory-scale system are listed in Table 5.7. As for the previous prototypes, thermocouples of type K with 0.5 mm of outer diameter were used as temperature sensors. In the catalyst brick, 4 sensors were equally distributed along the central axis. Additional measurement positions were at the heat exchanger's inflow and outflow ports and at both ends of the side channel between heat exchanger and catalyst.

As proposed in Section 4.3, the complete system was axially reversed in order to bring the catalyst closer to the exhaust gas source during heating mode. A pneumatically actuated exhaust flap (BOYSEN, [38]) was used to switch between two operating modes. With open flap the exhaust passes through the catalyst and the heat exchanger's straight outflow channels. If the flap is closed, the exhaust is redirected through an additional rectangular bypass channel (see Figure 5.15, bottom) to the sidewise inlet of the heat exchanger. From there, it passes through the inflow chan-

Table 5.7.: Geometric parameters of laboratory-scale sequential system.

Heat exchanger length	$[m]$	0.3
Ceramic TWC length	$[m]$	0.088
Cross-sectional area	$[m^2]$	0.0025
Channel height heat exch.	$[mm]$	1.5
Sheet thickness wall heat exch.	$[\mu m]$	110
Wall thickness TWC	$[mil]$	6
Sheet thickness spacer	$[\mu m]$	50
Cell density heat exch.	$[cpsi]$	385
Cell density TWC	$[cpsi]$	400

nels, the (upper) side channel and flows back through catalyst and outflow channels of the heat exchanger. A second flap is used to completely bypass the sequential system during start-up of the gas burner. In Section A.1.3, a detailed description of the experimental facilities is given.

5.3.2. Transient cold start experiments

With the sequential system a series of cold start experiments was performed. In each case, the burner flow rate was adjusted to a certain constant value until reaching stationary conditions in terms of exhaust gas temperature ($T_{exh.} \approx 180°C$) and composition. During this initial phase, the exhaust was fed through the bypass branch in order to protect the sequential system from temperature pikes due to sudden jumps in hydrocarbon concentration. Before starting the actual heating sequence, dosage of methane and CO was activated in order to attain a slightly lean exhaust gas composition (i.e. $\lambda = 1.01$). Then, an additional constant methane flow was superimposed with a three-way valve periodically alternating the feed position between exhaust hood and reactor inflow. As a result, sharp transitions between the lean base flow (i.e. additional methane flow to hood) and rich conditions (i.e. additional methane flow mixed to exhaust stream) with λ varying between 1.01 and 0.99 and switching frequencies of 1 Hz were obtained. Concentration of CO was constant at 1.1% while CH_4 varied due to the oscillating strategy between 2500 ppm and 2800 ppm. These values correspond to a total adiabatic temperature rise of 160 - 170 K. In an engine application, similar values could be obtained by late fuel injection leading to incomplete combustion in the cylinders. In combination with late ignition for increased exhaust temperature this is a common strategy for catalyst heating during cold start of gasoline engines. After sufficient catalyst heating, the bypass flap was closed, the electric heater was switched off and the system was allowed to stabilize in countercurrent operating mode. When the maximum temperature measured in the catalyst brick reached 600°C, the experiment was stopped in order to prevent the heat exchanger from excessive temperatures. The resulting transient evolutions of temperature and conversion are depicted in Figure 5.17 and Figure 5.18 for three different values of $GHSV$, respectively. As soon as catalytic conversion starts, the average temperature in the catalyst brick (black solid lines in Figure 5.17) rises above the feed temperature (black dashed lines in Figure 5.17) which is equivalent to the temperature measured in the head chamber (i.e. position "3" in Figure 5.16) during the heating phase. In case of 450 seconds of initial heating (left columns in Figure 5.17 and Figure 5.18), conversion of CO and CH_4 reaches values above 85% towards the end of bypass operation and the mean catalyst temperature is well above the light-off range. When the flap is closed and the electric heater switched off, the temperature in the head chamber drops immediately to some extent but it obviously remains above the level required for CH_4 light-off. On the contrary, 300 s of initial heating (right columns in Figure 5.17 and Figure 5.18) are not sufficient and CH_4 conversion considerably decreases after passing to countercurrent operation without further heating support.

In fact, the characteristic behavior observed in these experiments is very similar to the simulation results discussed in Section 4.3.2.2. There, the side channel between heat exchanger and catalyst was held responsible for causing a considerable decrease of exhaust temperature during the initial phase of countercurrent operation. During catalyst heating with the exhaust passing through the ceramic monolith and the heat exchanger's outflow channels, the side channel's walls remain relatively cold. When the flap is closed, the exhaust passes through the inflow channels of the heat exchanger and the side channel before reaching the catalyst. A temperature drop occurs until the side channel is completely heated up. Additionally, heat losses are expected to play a major role in case of the small laboratory-scale systems. Some leakage of cold exhaust might take place through gaps of the closed bypass flap since these devices were originally developed for acoustic purposes in the exhaust gas system where gas tightness is no prerequisite. However, the results are somehow contradictory since the temperature drop decreases with increasing throughput. In fact, higher flow rates lead to increased pressure drop over the closed flap. The leakage flow should therefore increase accordingly, leading to an equal cooling effect independent of the total flow rate.

In order to quantify the effect of the cold side channel walls on the exhaust, the temperature difference between its inlet (i.e. outlet temperature of the heat exchanger's inflow channels) and outlet was measured. In Figure 5.19 the resulting values at the three observed total flow rates are depicted.

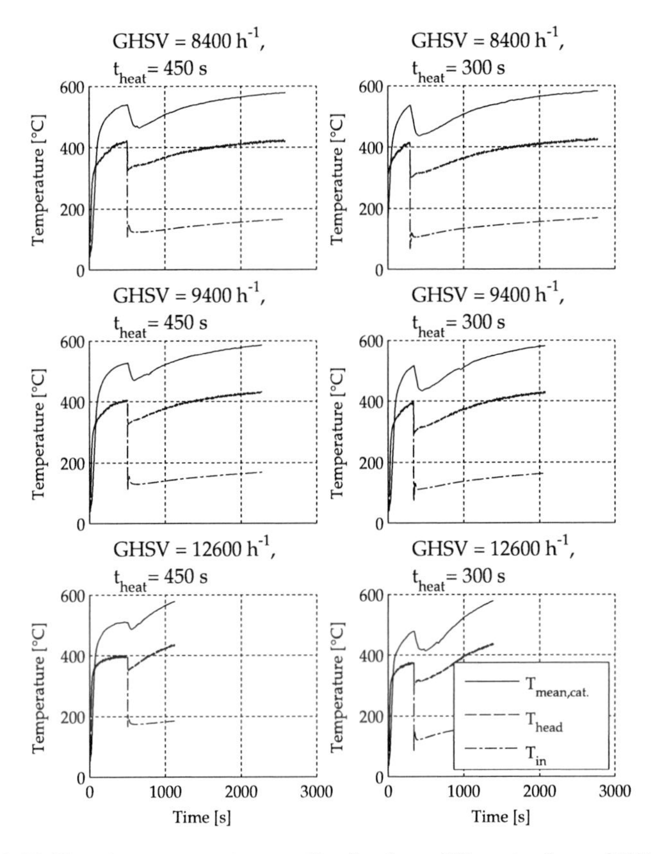

Figure 5.17.: Transient temperature profiles for three different values of *GHSV*. Left column: 450 s heating phase, right column: 300 s heating phase. $T_{mean,cat}$ denotes average of the 4 thermocouples in catalyst, T_{head} is measured in the head chamber (i.e. position "3" in Figure 5.16, bottom). T_{in} denotes feed temperature measured at position "3" during heating (with activated heater) and at position "1" during countercurrent operation.

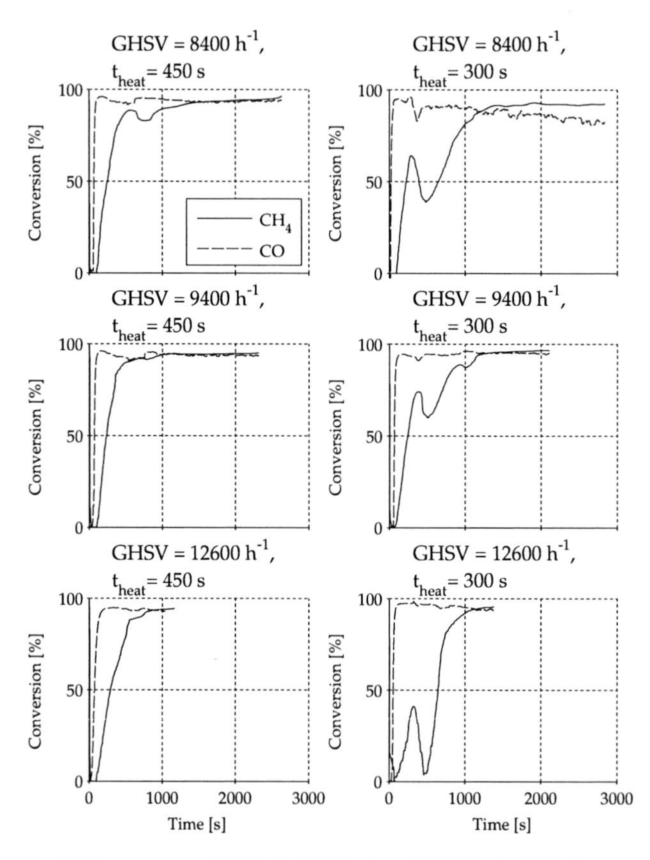

Figure 5.18.: Sliding average of CH_4 (solid line) and CO (dashed line) conversion corresponding to temperature results shown in Figure 5.17. Left column: 450 s heating phase, right column: 300 s heating phase

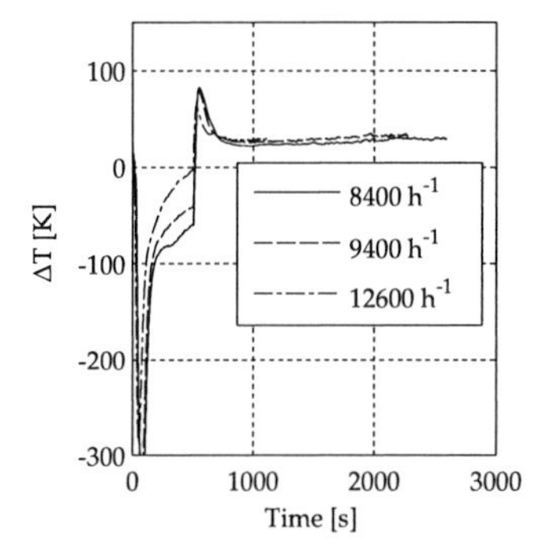

Figure 5.19.: Temperature difference between inlet and outlet of side channel for three different values of $GHSV$ and 450 s of initial heating.

Obviously, shortly after starting the heating sequence negative values of ΔT are obtained since the temperature front has not yet penetrated the catalyst brick. The side channel outlet temperature is therefore higher than the inlet one. However, when the bypass flap is closed the temperature of the gas entering the side channel from the already preheated inflow channels of the heat exchanger is higher than the temperature obtained at the side channel's outlet where a sudden temperature drop occurs when the bypass flap is closed (see Fig. 5.17). When the channel walls are completely heated up (after ~200 s), the temperature difference over the side channel reaches a stationary level (~30-40 K). This is due to heat losses over the side channel's walls.

5.3.3. Stationary experiments

A series of steady-state experiments was performed with the sequential system in order to compare its performance to the previous, integrated prototypes. The feed gas composition differed slightly from the transient experiments with the CO concentration decreased to 0.5%. Since λ was kept equal for optimal catalyst performance (i.e. $0.99 - 1.01$ with 1 Hz switching frequency), the CH_4 concentration was increased to $3200 - 4500$ ppm.

These values correspond to a total adiabatic temperature rise of $\sim$ 120 K. The results were obtained subsequent to the heat-up tests shown in the previous section at three different total flow rates. Hence, before gathering the stationary data at each of the three operating points, the CO concentration was adjusted and the system

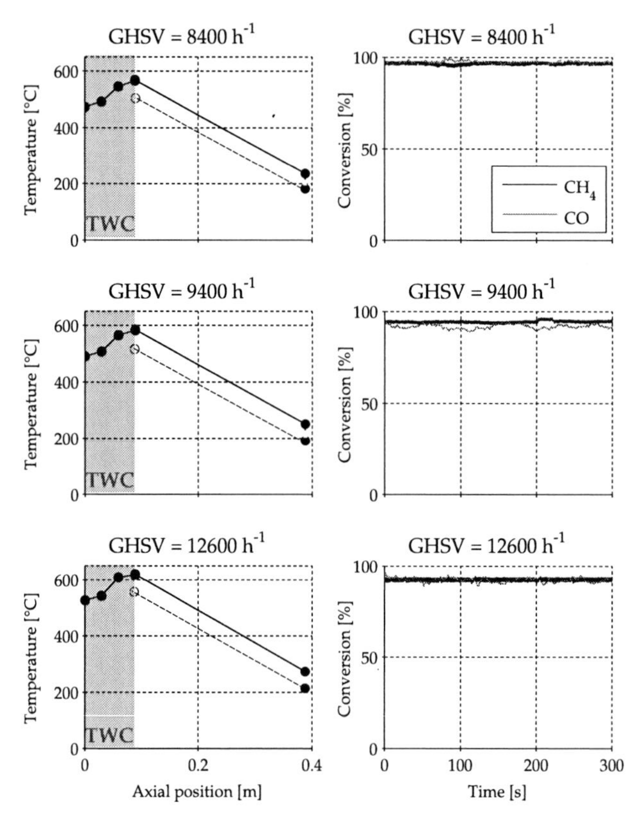

Figure 5.20.: Measured, stationary temperature profiles obtained with sequential system at three different values of *GHSV* (left column). Estimated temperature values in inflow channels of heat exchanger are shown as dashed lines/symbols. Right column: conversion of CO and CH_4.

was allowed to converge. In Figure 5.20, the resulting axial temperature profiles are shown together with conversion of CO and CH_4 which was recorded over a time span of 300 s. It is important to note, that in this case the plots' layout correspond to the "normal mode" of Figure 5.13 with the feed gas entering the heat exchanger from right to left.

Similar to the axial profiles obtained with the integrated prototypes without heat loss compensation (Figure 5.4 and Figure 5.8), a pronounced drop of temperature is observed between outlet of the heat exchanger's inflow channels (dashed symbols at right border of shaded area in Figure 5.13) and catalyst inlet at $z = 0$ m. Unlike in case of the previous prototypes, this temperature decrease is hardly affected by increased total flow rates. One reason for this behavior is the side channel between heat exchanger and catalyst. As shown in Figure 5.19, heat losses over the channel's walls lead to an almost constant temperature decrease of $\sim$ 30 K under stationary conditions. Additionally, the closed exhaust flap, which faces the catalyst brick, is constantly cooled by the relatively cold exhaust passing along its back side. It therefore acts as a strong heat sink at the monolith catalyst entrance. Together with heat losses over the head chamber's walls these effects lead to a total temperature drop of $\sim$ 100 K between heat exchanger and catalyst during stationary conditions. As a consequence, the amplification factor reached with the sequential system hardly exceeds 3 which is clearly inferior to the results obtained with the integrated systems. This underlines the demand for optimizing the design especially of the side channel / head chamber part of the sequential system in oder to decrease heat losses. Nevertheless, the system's performance is sufficient to maintain almost full conversion of CH_4 in spite of feed temperatures significantly below methane light-off.

5.4. Conclusions

A new design approach based on a vacuum-brazed, stainless steel heat exchanger was presented. The main goal of this prototype was to simplify the manufacturing process of the heat exchanger core compared to a previous concept while maintaining its amplification performance. At laboratory scale and under stationary conditions, very similar results as with a previously tested folded-sheet prototype of equal size were obtained. Backpressure was reduced by applying spacer structures with different cell densities in the inert and coated part of the heat exchanger.

At elevated space velocities, methane conversion was found to be increasingly limited by the relatively low specific surface in the catalyst coated part compared to the folded-sheet prototype. Moreover, the weight of the brazed heat exchanger core is $\sim$ 50% higher due to the increased material thickness and the braze filler material applied. This is a severe disadvantage for transient operation as demonstrated by simulation results in Section 4.2.

As a result, a sequential heat-integrated system according to the design concept introduced in Section 4.3 was built up for testing at laboratory scale. An electric heater was applied for cold start and pneumatically actuated exhaust flaps were used to redirect the exhaust stream in the respective operating mode. The results showed that the sequential system allows to quickly reach operating conditions suitable for safe catalytic conversion of CH_4. After transition to countercurrent operation the temperature level can be maintained and even increased despite considerably lower feed temperatures which would lead to immediate extinction of CH_4 conversion in a standard system without heat exchanger. Yet, as predicted by simulation results shown in Section 4.3.2.2, immediately after closing the bypass flap a pronounced decrease of average catalyst temperature was observed. This behavior was found to be predominantly due to the side channel between heat exchanger and catalyst which remains relatively cold during initial heating. When the bypass flap is closed and the system operated in countercurrent mode, the cold channel walls act as heat sink on the passing exhaust. Furthermore, heat loss over the channel's walls, the exhaust flap and the head chamber's wall surfaces impinge on the system's stationary performance. Therefore, further optimization of the design approach should concentrate on minimizing heat losses especially in this part of the system.

Chapter 6.

Directions for future work

Concepts for CNG exhaust

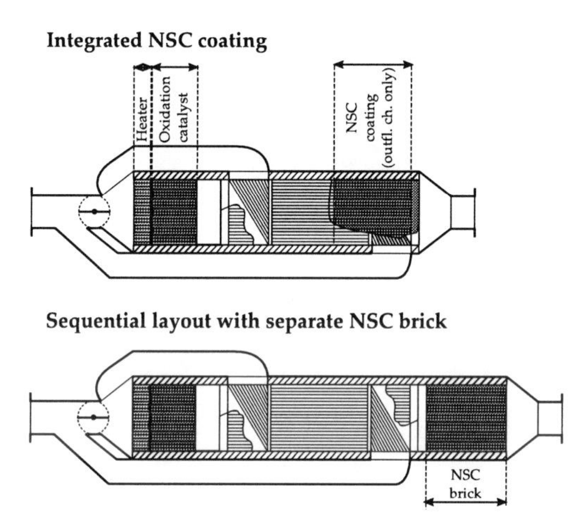

Figure 6.1.: Sequential concepts with integrated NSC coating (top) and separate ceramic bick (bottom).

In this work, the potential of heat-integrated exhaust aftertreatment systems has been demonstrated for CNG engines running under stoichiometric conditions. Lean-burn concepts allow for even increased engine efficiency [20] and therefore represent the next stage of development. The exhaust of these engines however contains considerable amounts of NO_x and hydrocarbon (HC) emissions at significantly decreased exhaust temperatures [20, 73]. While the removal of HC emissions can be achieved by a combination of oxidation catalyst and heat exchanger, the removal of

NO_x emissions is challenging and requires additional purification stages. A tantalizing option would be to apply a combined NO_x storage / oxidation catalyst material at the heat exchanger's hot end. However, due to the amplification effect of the countercurrent heat exchanger, the temperatures reached in this part will lead to severe thermal damaging, especially of the storage catalyst. Alternatively, the NO_x storage catalyst (NSC) could be integrated into the heat exchanger's outflow channels at the cold end (Figure 6.1, top). Yet, in this case the available active surface would be rather small due to the demand for wide channels in order to minimize backpressure and the fact that only half of the heat exchanger's channel volume is utilized. Therefore, a concept with separate NSC brick downstream of the heat exchanger, which is completely passed by the exhaust, is regarded more favorable (Figure 6.1, bottom). By adding an AdBlue® doser and replacing the NSC brick with a stage for selective catalytic reduction (SCR), another layout for NO_x removal could be realized. This underlines the relatively easy adaption of the sequential system to different purification scenarios.

Integrated, multi-functional converter concept for diesel exhaust

In case of diesel engines, exhaust aftertreatment becomes more challenging since particulate matter has to be removed in addition to HC and NO_x. In a previous publication, a fully integrated concept, combining all of these functionalities in one device, was proposed [5].

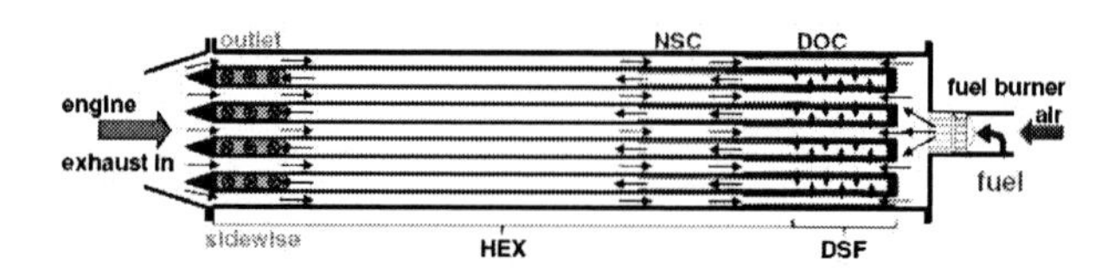

Figure 6.2.: Integrated system for purification of diesel exhaust based on a ceramic heat exchanger containing different purification stages [5].

In Figure 6.2, a schematic of this system is depicted. The heat exchanger is based on a ceramic monolith with every second channel row (i.e. the outflow channels) sealed at one end. In a zone located at the right end, the wall is porous and serves as diesel soot filter (DSF). Over the remaining length of the heat exchanger, the walls are sealed in order to prevent the unpurified exhaust from bypassing the catalytic part. Within and upstream of the filter zone, a diesel oxidation catalyst (DOC) is deposited at the channel walls of both the inflow and outflow channels. Downstream of the DOC/DSF part, the channel walls are coated with a NSC material over a certain length. Due to heat integration, the DSF zone can be maintained above 300°C for most of the operating time. This temperature range is sufficient to initiate continuous soot oxidation by NO_2, which is formed by NO and O_2 at the DOC. As products of this mechanism, which is commonly known as CRT® reaction [14, 15], CO and NO are formed. The former is further oxidized at the DOC, yielding CO_2, while the latter needs to be stored in the downstream NSC zone. From time to time, the exhaust gas composition is switched to rich conditions in order to regenerate the NO_x trap. As cold start support and for generating CO/H_2-rich gas mixtures, a (catalytic) fuel burner was envisaged at the heat exchanger's head end.

While this system layout is clearly beneficial in terms of energy efficiency and compact packaging, the ceramic heat exchanger with multiple coatings also exhibits several disadvantages. First of all, the application of different washcoat materials within well-defined zones is rather difficult to achieve. As pointed out above, the available volume of the NSC stage is rather small since only the outflow channels are coated and the channels have to be sufficiently wide to avoid excessive backpressure. Furthermore, reliable sealing of the channel walls within the heat exchanger zone requires additional manufacturing effort. Finally, after having gained some experience with these systems, operation of the fuel burner under highly dynamic operating conditions is expected to be very challenging.

Sequential system for diesel exhaust

As a consequence of the various disadvantages of the integrated approach described in the previous section, a modified version of the new sequential system presented in this work can be used for the treatment of diesel exhaust. To this end, the various different purification stages are combined with an inert heat exchanger and a bypass/heater system for accelerated heat-up. In Figure 6.3, schematic pictures of the resulting system are depicted [19].

During cold start (Figure 6.3, top), the exhaust flap is opened, allowing the particle-loaded raw exhaust to bypass the heat exchanger's inflow channels and directly enter the catalyst stages. This layout is chosen in order to avoid deposition of large soot agglomerates in the heat exchanger during engine start-up. Before entering the catalyst stages, the exhaust temperature can be further increased by an electric heater, as proposed in this work. Subsequently, the exhaust passes through a wall-flow ceramic DPF which is coated with a DOC. Then, the flow direction is reversed and the exhaust enters the two side channels which contain SCR catalysts for NO_x reduction. Before entering this purification stage, AdBlue® is dosed in the head chamber, which, by hydrolysis, delivers the ammonia required for NO_x reduction in the SCR stage. After leaving the side channels, the purified exhaust passes through the heat exchanger's outflow channels. Both inlet and outlet occur over sidewise ports. As

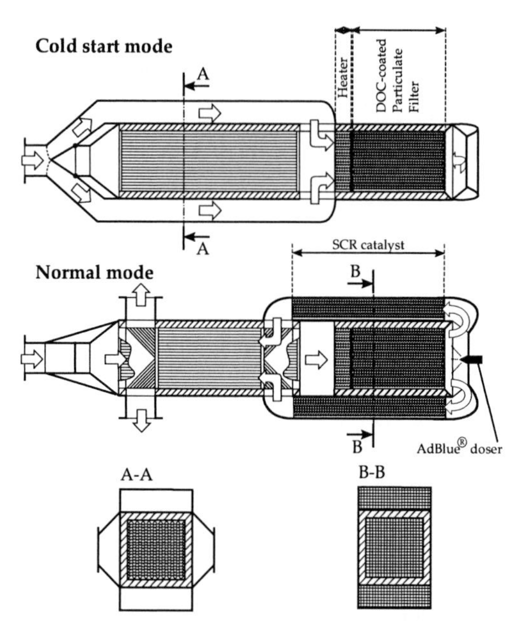

Figure 6.3.: Concept of sequential system for purification of diesel exhaust. Top: sectional view showing the exhaust flow during cold start operating mode. Center: sectional view, rotated about central axis by 90°, showing flow configuration during normal operating mode. Bottom: additional sectional views, highlighting the position of the different side channels.

soon as the catalyst stages are sufficiently heated up, the bypass flap is closed. Now, the raw exhaust passes first through the inflow channels of the heat exchanger before entering the catalyst stages (Figure 6.3, center). Due to the straight, parallel inflow channels without flow deflection, the danger of clogging by soot deposits is minimized. The outflow path is equal to cold start operating mode. As discussed in this work, the bypass flap can be re-opened when the exhaust temperature is sufficiently high for reliable operation of the catalyst stages without the heat exchanger's amplification effect. This leads to decreased backpressure especially during engine high-load conditions and prevents overheating of the catalyst stages.

The sequential layout of this multifunctional exhaust purification system has several advantages over the integrated approach discussed in the previous section and over the standard solution without heat exchanger. For the catalyst stages, standard ceramic substrates can be applied. This allows for a specific adaption of brick volumes and cell densities to the respective exhaust conditions. For the heat exchanger, a robust metallic layout, as described in this work, can be applied without the need for complex sealing and coating of the channel walls. Finally, the combination of bypass flap and electric heater allows for excellent controllability during system cold start. Compared to the standard system without heat exchanger, lower brick volumes can be applied since the catalyst stages are operated in an optimal temperature range. This allows for some reduction of cost (i.e. less precious metal required) and backpressure. Moreover, since the DPF is operated within a range, where continuous regeneration can take place, the average soot loading should remain relatively low, leading to reduced backpressure.

In the standard system, the temperature before SCR oftentimes remains below the range required for reliable evaporation of AdBlue® and optimal performance of the SCR catalyst. This is especially severe in case of urban buses, which typically operate with low to medium engine loads, interrupted by short phases of high load during acceleration. As a consequence, depositions are formed around the doser and even at the inlet of the SCR catalyst which can cause a complete failure of the system. In the sequential, heat-integrated system however, the temperature of all catalyst stages can be maintained at sufficiently high levels even during long phases of low engine load.

Bibliography

[1] BAEHR, H. D., AND STEPHAN, K. *Wärme- und Stoffübertragung*, 7th ed. Berlin: Springer, 2010.

[2] BAZ, J. Experimentelle Untersuchung des Kaltstartverhaltens von wärmeintegrierten Abgasnachbehandlungssystemen. B.Sc. thesis, Universität Stuttgart, 2011.

[3] BEKYAROVA, E., FORNASIERO, P., KASPAR, J., AND GRAZIANI, M. CO oxidation on $Pd/CeO_2 - ZrO_2$ catalysts. *Catalysis Today 45*, 1-4 (1998), 179 – 183.

[4] BERGHEN, F. V. *Constrained, non-linear, derivative-free, parallel optimization of continuous, high computing load, noisy objective functions*. PhD thesis, Université Libre de Bruxelles, 2004.

[5] BERNNAT, J., RINK, M., DANNER, T., TUTTLIES, U., NIEKEN, U., AND EIGENBERGER, G. Heat-integrated concepts for automotive exhaust purification. *Topics in Catalysis 52* (2009), 2052–2057.

[6] BHATIA, S. K. Analysis of Catalytic Reactor Operation with Periodic Flow Reversal. *Chemical Engineering Science 46(1)* (1991), 361.

[7] BOUCHER, O., FRIEDLINGSTEIN, P., COLLINS, B., AND SHINE, K. P. The indirect global warming potential and global temperature change potential due to methane oxidation. *Environmental Research Letters 4* (2009), 5pp.

[8] BOUNECHADA, D. *Strategies for the enhancement of low-temperature catalytic oxidation of methane emissions*. PhD thesis, Politecnico di Milano, 2012.

[9] BOUNECHADA, D., GROPPI, G., FORZATTI, P., KALLINEN, K., AND KINNUNEN, T. Effect of periodic lean/rich switch on methane conversion over a Ce-Zr promoted Pd-Rh/Al_2O_3 catalyst in the exhaust of natural gas vehicles. *Applied Catalysis B: Environmental 119-120* (2012), 91–99.

[10] BRINKMEIER, C. *Automotive Three-Way Exhaust Aftertreatment under Transient Conditions - Measurements, Modeling and Simulation*. PhD thesis, Universität Stuttgart, 2006.

[11] BURCH, R., CRITTLE, D. J., AND HAYES, M. J. C-H bond activation in hydrocarbon oxidation on heterogeneous catalysts. *Catalysis Today 47* (1999), 229–234.

[12] BURCH, R., LOADER, P. K., AND URBANO, F. J. Some aspects of hydrocarbon activation on platinum group metal combustion catalysts. *Catalysis Today 27* (1996), 243–248.

[13] CASTELLAZZI, P., GROPPI, G., FORZATTI, P., BAYLET, A., MARÉCOT, P., AND DUPREZ, D. Role of Pd loading and dispersion on redox behaviour and CH_4 combustion activity of Al_2O_3 supported catalysts. *Catalysis Today 155*, 1-2 (2010), 18 – 26.

[14] COOPER, B., JUNG, H., AND THOSS, J. Treatment of diesel exhaust gases. US Patent 4,902,487A, Johnson Matthey, 1988.

[15] COOPER, B. J., AND THOSS, J. E. Role of NO in Diesel Particulate Emission Control. *SAE paper 890404* (1989), 171–183.

[16] COTTRELL, F. G. Purifying gases and apparatus therefore. US-Patent 2,121,733, 1938.

[17] DIMOPOULOS, P., BACH, C., SOLTIC, P., AND BOULOUCHOS, K. Hydrogen-natural gas blends fuelling passenger car engines: Combustion, emissions and well-to-wheels assessment. *International Journal of Hydrogen Energy 33* (2008), 7224–7236.

[18] EIGENBERGER, G., AND NIEKEN, U. Catalytic cleaning of polluted air: Reaction engineering problems and new solutions. *International Chemical Engineering 34(1)* (1994), 4–16.

[19] EIGENBERGER, G., NIEKEN, U., RINK, M., AND MATSCHKE, S. Verfahren und Vorrichtung zur Abgasreinigung mit optionaler Wärmerückgewinnung für Verbrennungskraftmaschinen. German Patent DE10201110976A1, 2011.

[20] EINEWALL, P., TUNESTÅL, P., AND JOHANSSON, B. Lean Burn Natural Gas Operation vs. Stoichiometric Operation with EGR and a Three Way Catalyst. *SAE paper 2005-01-0250* (2005).

[21] ENGERER, H., AND HORN, M. Natural gas vehicles: An option for Europe. *Energy Policy 38* (2010), 1017–1029.

[22] FORSTER, P., RAMASWAMY, V., ARTAXO, P., BERNTSEN, T., BETTS, R., FAHEY, D. W., HAYWOOD, J., LEAN, J., LOWE, D. C., MYHRE, G., NGANGA, J., PRINN, R., RAGA, G., SCHULZ, M., AND DORLAND, R. V. *Climate Change 2007: The*

Physical Science Basis. Contribution of Working Group I to the Fourth Assessment Report of the Intergovernmental Panel on Climate Change. Cambridge University Press, Cambridge, United Kingdom and New York, NY, USA, 2007, ch. Changes in Atmospheric Constituents and in Radiative Forcing, pp. 129–234.

[23] FRAUHAMMER, J. *Ein neues Gegenstrom-Reaktorkonzept für endotherme Hochtemperaturreaktionen.* PhD thesis, Universität Stuttgart, 2003.

[24] FREIDRICH, G., GAISER, G., EIGENBERGER, G., OPFERKUCH, F., AND KOLIOS, G. Kompakter Reaktor für katalytische Reaktionen mit integriertem Wärmerücktausch. European Patent EP0885653B1, 2003.

[25] FRIEDRICH, G., KOLIOS, G., SCHMEISSER, V., TUTTLIES, U., OPFERKUCH, F., AND EIGENBERGER, G. Verfahren und Vorrichtung zur Reinigung von Abgasen. Patent WO2004/099577, 2004.

[26] GÉLIN, P., URFELS, L., PRIMET, M., AND TENNA, E. Complete oxidation of methane at low temperature over Pt and Pd catalysts for the abatement of lean-burn natural gas fuelled vehicles emissions: influence of water and sulphur containing compounds. *Catalysis Today 83* (2003), 45–57.

[27] GNIELINSKI, V. Wärmeübertragung bei der Strömung durch Rohre. *VDI-Wärmeatlas 10* (2006), Ga1–Ga9.

[28] GOLUBITSKY, M., AND SCHAEFFER, D. G. *Singularities and groups in bifurcation theory, Vol. I.* Springer-Verlag, New York, 1985.

[29] GRITSCH, A. *Wärmeintegrierte Reaktorkonzepte für katalytische Hochtemperatur-Synthesen am Beispiel der dezentralen Dampfreformierung von Methan.* PhD thesis, Universität Stuttgart, 2008.

[30] HEKKERT, M. P., HENDRIKS, F. H. J. F., FAAIJ, A. P. C., AND NEELIS, M. L. Natural gas as an alternative to crude oil in automotive fuel chains, well-to-weel analysis and transition strategy development. *Energy Policy 33* (2005), 579–594.

[31] JOBSON, E., AND HEED, B. Catalytic purification device. European Patent 1016777B1, 1999.

[32] JONES, E., OLIPHANT, T., PETERSON, P., ET AL. SciPy: Open source scientific tools for Python, 2001–.

[33] K.FUJIMOTO, RIBEIRO, F. H., AVALOS-BORJA, M., AND IGLESIA, E. Structure and Reactivity of of PdOx/ZrO2 Catalysts for Methane Oxidation at Low Temperatures. *Journal of Catalysis 179* (1998), 431–442.

[34] KINNUNEN, N., SUVANTO, M., MORENO, M., SAVIMÄKI, A., KALLINEN, K., KINNUNEN, T.-J., AND PAKKANEN, T. Methane oxidation on alumina supported palladium catalysts: Effect of Pd precursor and solvent. *Applied Catalysis A: General 370* (2009), 78–87.

[35] KINNUNEN, N. M., HIRVI, J. T., VENÄLÄINEN, T., SUVANTO, M., AND PAKKANEN, T. A. Procedure to tailor activity of methane combustion catalysts: Relation between Pd/PdO_x active sites and methane oxidation activity. *Applied catalysis A: General 397* (2011), 54–61.

[36] KLINGSTEDT, F., NEYESTANAKI, A. K., BYGGNINGSBACKA, R., LINDFORS, L., LUNDÉN, M., PETERSSON, M., TENGSTRÖM, P., OLLONQVIST, T., AND VÄYRYNEN, J. Palladium based catalysts for exhaust aftertreatment of natural gas powered vehicles and biofuel combustion. *Applied Catalysis A: General 209* (2001), 301–316.

[37] KOLIOS, G., FRAUHAMMER, J., AND EIGENBERGER, G. Autothermal fixed-bed reactor concepts. *Chemical Engineering Science 55* (2000), 5945–5967.

[38] KORNHERR, H., STOCKINGER, K., AND WELLNER, F. Device for closing a fluid stream. European Patent 1503062, 2004.

[39] KRASNYK, M. *DIANA - An object-oriented tool for nonlinear analysis of chemical processes*. PhD thesis, Otto-von-Guericke-Universität Magdeburg, 2008.

[40] KRASNYK, M., BONDAREVA, K., MILOKHOV, O., TEPLINSKIY, K., GINKEL, M., AND KIENLE, A. The ProMoT/DIANA simulation environment. In *16th European Symposium on Computer Aided Process Engineering and 9th International Symposium on Process Systems Engineering*, vol. 21 of *Computer Aided Chemical Engineering*. 2006, pp. 445 – 450.

[41] LINSTROM, P. J., AND MALLARD, W. G., Eds. *NIST Chemistry WebBook, NIST Standard Reference Database Number 69*. National Institute of Standards and Technology, Gaithersburg MD, 20899, http://webbook.nist.gov, retrieved 2008.

[42] LIQUEFIED PETROLEUM GASES; PROPANE, PROPENE, BUTANE, BUTENE AND THEIR MIXTURES; REQUIREMENTS. Norm DIN 51622, 1985.

[43] M. W. CHASE, JR. NIST-JANAF Thermochemical Tables, Fourth Edition. *J. Phys. Chem. Ref. Data, Monograph 9* (1998), 1–1951.

[44] MATROS, Y. S. *Catalytic Processes Under Unsteady-State Conditions*. Elsevier, 1989.

[45] MATROS, Y. S., AND BUNIMOVICH, G. A. Reverse-Flow Operation in Fixed Bed Catalytic Reactors. *Catalysis Reviews Science and Engineering 38(1)* (1996), 1–68.

[46] MAUS, W., BRÜCK, R., KONIECZNY, R., AND SCHEEDER, A. Der E-Kat als Thermomanagementlösung in modernen Fahrzeuganwendungen. *MTZ Motortechnische Zeitschrift 71* (2010), 340–346.

[47] MÜLLER, C. A., MACIEJEWSKI, M., KOEPPEL, R. A., AND BAIKER, A. Combustion of methane over palladium/zirconia: effect of Pd-particle size and role of lattice oxygen. *Catalysis Today 47* (1999), 245–252.

[48] NEYESTANAKI, A. K., KLINGSTEDT, F., SALMI, T., AND MURZIN, D. Deactivation of postcombustion catalysts, a review. *Fuel 83* (2004), 395–408.

[49] NIEKEN, U. *Abluftreinigung in katalytischen Festbettreaktoren bei periodischer Strömungsumkehr.* PhD thesis, Universität Stuttgart, 1993.

[50] NIEKEN, U. *Chemische Reaktionstechnik I, Vorlesungsskript.* Universität Stuttgart, 2004.

[51] NIEKEN, U., KOLIOS, G., AND EIGENBERGER, G. Control of the ignited steady state in autothermal fixed-bed reactors for catalytic conversion. *Chemical Engineering Science 49* (1994), 5507–5518.

[52] NIEKEN, U., KOLIOS, G., AND EIGENBERGER, G. Fixed-bed reactors with periodic flow reversal: experimental results for catalytic combustion. *Catalysis Today 20* (1994), 335–350.

[53] NIEKEN, U., KOLIOS, G., AND EIGENBERGER, G. Limiting Cases and Approximate Solutions for Fixed-Bed Reactors with Periodic Flow Reversal. *AIChE Journal 41*, 8 (1995), 1915–1925.

[54] OBUCHI, A., UCHISAWA, J., OHI, A., NANBA, T., AND NAKAYAMA, N. A catalytic diesel particulate filter with a heat recovery function. *Topics in Catalysis 42-43* (2007), 267–271.

[55] OH, S. H., MITCHELL, P. J., AND SIEWERT, R. M. Methane oxidation over alumina-supported noble metal catalysts with and without cerium additives. *Journal of Catalysis 132*, 2 (1991), 287 – 301.

[56] OPFERKUCH, F. *Autotherme Reaktoren zur Rückhaltung und Umwandlung von oxidierbaren Partikeln in Abgas.* PhD thesis, Universität Stuttgart, 2004.

[57] PECK, R. *Experimentelle Untersuchung und dynamische Simulation von Oxidationskatalysatoren und Diesel-Partikelfiltern.* PhD thesis, Universität Stuttgart, 2007.

[58] PELKMANS, L., AND DEBAL, P. Comparison of on-road emissions with emissions measured on chassis dynamometer test cycles. *Transportation Research, Part D 11* (2006), 233–241.

[59] PFAHL, U., SCHATZ, A., AND KONIECZNY, R. Advanced Exhaust Gas Thermal Management for Lowest Tailpipe Emissions - Combining Low Emission Engine and Electrically Heated Catalyst. *SAE paper 2012-01-1090* (2012).

[60] POWELL, M. UOBYQA: Unconstrained Optimization By Qadratic Approximation. Tech. Rep. DAMTP1997/NA12, Department of Applied Mathematics and Theoretical Physics, University of Cambridge, England, 1997.

[61] POWELL, M. J. D. *A direct search optimization method that models the objective and constraint functions by linear interpolation.* Kluwer Academic (Dordrecht), 1994.

[62] RATFISCH ANALYSENSYSTEME GMBH. *Richtig messen mit dem FID.* Device manual.

[63] REID, R., PRAUSNITZ, J., AND POLING, B. *The Properties of Gases & Liquids*, 5th ed. New York: McGraw-Hill, 2001.

[64] REISER, S. Untersuchung des stationären und dynamischen Betriebsverhaltens eines wärmeintegrierten Systems zur Abgasreinigung von monovalenten Erdgasfahrzeugen. Diploma thesis, Universität Stuttgart, 2010.

[65] RINK, M., EIGENBERGER, G., AND NIEKEN, U. Heat-integrated Exhaust Purification for Natural Gas Engines. *Chemie Ingenieur Technik 85* (2013), N.A.

[66] RINK, M., EIGENBERGER, G., NIEKEN, U., AND TUTTLIES, U. Optimization of a heat-integrated exhaust catalyst for CNG engines. *Catalysis Today 188* (2012), 113–120.

[67] SALOMONS, S., HAYES, R. E., AND AMD H. SAPOUNDJIEV, M. P. Flow reversal reactor for the catalytic combustion of lean methane mixtures. *Catalysis Today 83* (2003), 59–69.

[68] SAMUEL, S., AUSTIN, L., AND MORREY, D. Automotive test cycles for emission measurement and real-world emission levels - a review. *Proceedings of the Institution of Mechanical Engineers, Part D: Journal of Automobile Engineering 216* (2002), 555–564.

[69] SEMENOFF, N. Zur Theorie des Verbrennungsprozesses. *Zeitschrift für Physik A Hadrons and Nuclei 48* (1928), 571–582. 10.1007/BF01340021.

[70] SMOKERS, R., VERMEULEN, R., VAN MIEGHEM, R., GENSE, R., SKINNER, I., FERGUSSON, M., MACKAY, E., AND TEN BRINK, P. Review and analysis of the reduction potential and costs of technological and other measures to reduce co_2-emissions from passenger cars. Report to the European Commission, TNO Report 06.OR.PT.040.2/RSM, 2006.

[71] TSINOGLOU, D. N., DIMOPOULOS EGGENSCHWILER, P., THURNHEER, T., AND HOFER, P. A simplified model for natural-gas vehicle catalysts with honeycomb and foam substrates. In *Proceedings of the Institution of Mechanical Engineers, Part D: Journal of Automobile Engineering* (2009), vol. 223, Institution of Mechanical Engineers, SAGE, pp. 819–834.

[72] U.S. ENERGY INFORMATION ADMINISTRATION: NATURAL GAS. Website, March 2013. http://www.eia.gov/naturalgas.

[73] VARDE, K. S., PATRO, N., AND DROUILLARD, K. Lean Burn Natural Gas Fueled S.I. Engine and Exhaust Emissions. *SAE paper 952499* (1995), 177–185.

[74] VORTMEYER, D., AND SCHÄFER, R. J. Equivalence of One and Two-Phase Models for Heat Transfer Processes in Packed Beds. *Chemical Engineering Science 29* (1974), 485.

[75] WS WÄRMEPROZESSTECHNIK. *REKUMAT C 150*, Ver00154. Manual.

[76] WÜNNING, J. Flammenlose Oxidation von Brennstoff mit hochvorgewärmter Luft. *Chemie Ingenieur Technik 63* (1991), 1243–1245.

[77] WÜNNING, J. G. Flameless oxidation burners for heating strip lines. *Heat processing* (October 2003), 1–4.

[78] WÜNNING, J. G. Regenerative Burners for Heat Treating Furnaces. In *8th European Conference on Industrial Furnaces and Boilers* (2008).

[79] ZAHN, V. M. *Adiabatic Simulated Moving Bed Reactor - Principle, Nonlinear Analysis and Experimental Demonstration*. PhD thesis, Otto-von-Guericke-Universität Magdeburg, 2012.

[80] ZAHN, V. M., MANGOLD, M., KRASNYK, M., AND SEIDEL-MORGENSTERN, A. Theoretical Analysis of Heat Integration in a Periodically Operated Cascade of Catalytic Fixed-Bed Reactors. *Chemical Engineering Technology 32* (2009), 1326–1338.

[81] ZAHN, V. M., MANGOLD, M., AND SEIDEL-MORGENSTERN, A. Autothermal operation of an adiabatic simulated counter cuurrent reactor. *Chemical Engineering Science 65* (2010), 458–465.

[82] ZAHN, V. M., YI, C.-U., AND SEIDEL-MORGENSTERN, A. Analysis and demonstration of a control concept for a heat integrated simulated moving bed reactor. *Chemical Engineering Science 66* (2011), 4901–4912.

Appendix A.

Experimental Facilities

In this chapter, the test rigs built up and used for experimental evaluation of the laboratory-scale prototypes shown in Chapter 5 are presented. Additionally, details of the exhaust generator system applied for the test runs with stoichiometric gas composition (Sections 5.2.4 to 5.3) are given.

A.1. Test rigs for experimental evaluation of heat-exchanger prototypes

A.1.1. Experimental setup for stationary, fuel-lean conditions

As base flow, pressurized air was dosed at ambient temperature with a mass flow controller (*Bronkhorst EL-Flow, Type: F-203-FA-44-V, Range: 0-30* Nm^3/h) and fed through an arrangement of three consecutive electric heaters with a maximum total power output of 18 kW (i.e. "Heater" in Fig. A.1). Due to limited residence time in each heating stage and/or limited maximum temperature of the heating cartridges, the air flow is heated gradually over the three stages with the controller of the final one adjusted to the actual feed setpoint temperature. For temperature control, three two-point controllers with manual setpoint input (*Horst HT 30*) were applied. Before entering the heat exchanger, the hydrogen (during system heat-up) and methane flux was added to the base flow. As for air, the fluxes were dosed with mass flow controllers (*Bronkhorst EL-Flow, Type: F-201C-RAD-33-K, Range: 0-5 Nl/min*). Thermocouples of type K with outer diameter of 0.5 mm (*Omega*) were inserted in the center channels of the heat exchanger from the coated end. Absolute pressure was recorded with three sensors (*Halstrup-Walcher, Range: 0-250 mbar* at inflow, half-way and outflow position. Gas samples were taken at the heat exchanger's inlet and outlet with a three-way valve for switching to either position. The sampling gas was fed through heated hoses (*Horst H 12/13*) and a cooler/condenser unit to the gas analyser. Since in this case methane was the only pollutant to be converted, a flame ionization detector (FID, *Ratfisch RS-53T*) was applied. Before each test run, the system was calibrated with a calibration gas mixture (5000 ppm CH_4 in synthetic air).

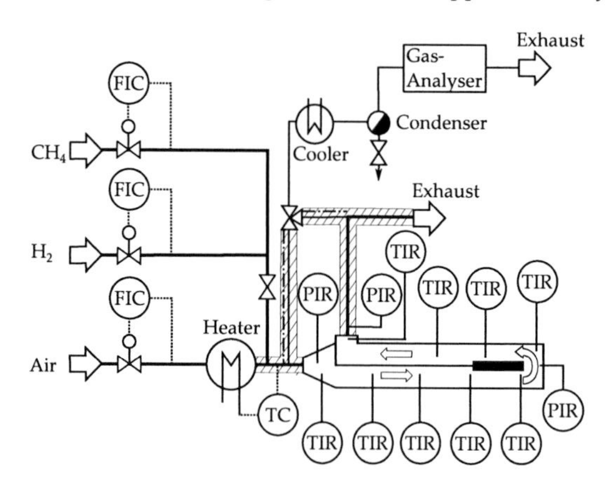

Figure A.1.: Test rig for stationary, fuel-lean conditions

The results of the initial stationary experiments conducted with both integrated heat-exchanger prototypes (see Section 5.1.2 and Section 5.2.2) were obtained with the same test rig whose flow chart is depicted in Figure A.1.

For test rig automation and real-time control, a *Gantner e.pac* system was used. This device allows to read measurement values from the different sensors and devices listed above via analog-in interface modules (*Gantner e.bloxx A4-1TC* for thermocouples and *Gantner e.bloxx A3-1* for general analog-in) which are all attached to a fast RS485 bus. Setpoints for the massflow controllers are passed to the respective device via analog-out interfaces (*Gantner e.bloxx A9-1*). On the e.pac, the measurement program is executed and collected data is stored. A TCP/IP connection to a Microsoft Windows PC allows for online readout of measurement values and input of setpoint updates. Moreover, a program editor is available in order to build and test the programs on the PC in offline mode prior to upload and execution on the e.pac. For program set-up, various pre-defined program modules are available which can be connected in a GUI-based editor.

A.1.2. Experimental setup for stationary, stoichiometric conditions

As described in Section 5.2.4, the operation of the Pd-based three-way catalyst under fuel-lean conditions led to a gradual decrease of CH_4 conversion efficiency. Besides mechanical damage of the washcoat, which occurred especially during manufacturing of the folded-sheet prototype, the complete oxidation of the precious metal was assumed to be the major issue leading to continuous deactivation. In order to reduce the catalyst, fuel-rich gas composition was required. Synthetic exhaust was not an option since the required fluxes were too high for a bottle-based gas supply. Consequently, a gas burner was applied as exhaust gas generator and the test rig was modified accordingly. The resulting setup is depicted in Figure A.2. The exhaust

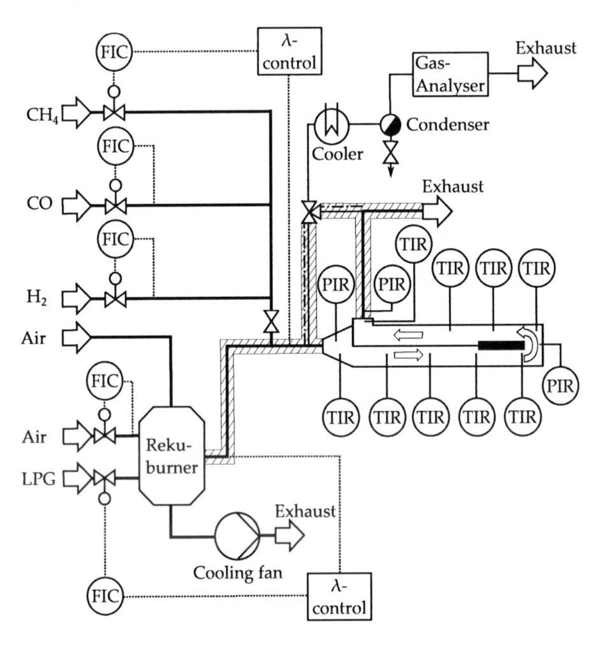

Figure A.2.: Test rig for stationary, near-stoichiometric conditions

gas generator ("Reku-burner" in the flow chart) is described in more detail in Section A.2 where also the composition of the raw exhaust is shown. Since Natural Gas was not available with the required pressure, Liquefied Petroleum Gas (LPG) [42] was used as fuel and dosed with an additional massflow controller (*MKS, Type: 1559AX-050L-SUS100, Range: 0-50 Nl/min*). Since LPG is usually the waste gas of oil refining processes, its composition is not exactly specified. Hence, in order to maintain a precise air-fuel ratio in the exhaust, the fuel mass flux was automatically adapted to the respective gas mixture based on the measurement signal of a wideband lambda sensor (*BOSCH LSU4.2 attached to ETAS LA4*) placed downstream of the burner. As air supply for the burner, pressurized air was used and dosed with the massflow controller already used during previous testing with fuel-lean conditions. Downstream of the burner, the exhaust gas composition could be further changed by addition of hydrogen and CO. For dosage of the latter component, a new massflow controller was added to the test rig (*Bronkhorst EL-Flow, Type: F-201C-FA-33-V, Range: 0-15 Nl/min*). In order to precisely specify the air-fuel ratio at heat-exchanger inlet, a second lambda-control loop was installed. As for the exhaust generator, a wideband lambda sensor was used to automatically control the methane feed based on a given setpoint value. For gas analysis, new sensors for CO, CO_2 (non-dispersive IR) and O_2 (electrochemical), combined in one device (*ABB Advance Optima*), were added. Since the burner was operated with excess air, no hydrocarbons were contained in the raw exhaust of the burner. Hence, the original approach of a FID for analysis of CH_4 content in the exhaust was retained. However, the originally applied calibration gas for the FID with high oxygen content was replaced by a mixture between methane and nitrogen (i.e. 5000 ppm CH_4 in N_2) due to a cross sensitivity of oxygen on the FID measurement [62]. The remaining test rig (i.e. sensor equipment of heat exchanger) was adopted from the previous setup for fuel-lean experiments.

A.1.3. Experimental setup for transient, stoichiometric conditions

For transient testing of the sequential heat-integrated system (results shown in Section 5.3), the test rig was considerably extended. The final setup is depicted in Figure A.3. In order to allow for oscillating air-fuel ratios in the exhaust, an additional mass flow controller (*MKS 1179AX14CM1BV, Range: 0-10 Sl/min*) for methane was added and coupled with a three-way valve allowing for two different feed positions. Initially, slightly lean (e.g. λ=1.01), stationary conditions were obtained by a constant flux of methane (dosed by the first mass flow controller) and a constant level of CO. The second methane flux was redirected into the exhaust line by switching the three-way valve to the first position. By changing over to the other valve position, the second methane flux was mixed with the steady methane/CO flux resulting in a sharp jump to rich conditions (e.g. λ=0.99). Timing of valve actuation and automatic calculation of the required second methane flow were realized by an extension of the previously applied program for stationary Lambda control.

For actuation of the two pneumatic exhaust flaps (V1 and V2 in Fig. A.3), an electric vacuum pump was used which automatically maintained a certain low-pressure level in a small reservoir vessel. For open flap position, the connection between the respective actuator and vacuum vessel is opened by switching a magnetic valve. By exposing the actuator to ambient pressure (i.e. magnetic valve to other position) the flap closes.

As power supply for the electric EMICAT heater, an automotive battery (12 V / 65 Ah) was applied and connected to a quick-charger. As switches for the high current required for heater operation (> 100 A), two relays connected in parallel were used.

More details regarding the electric equipment of the test rig are documented elsewhere [2].

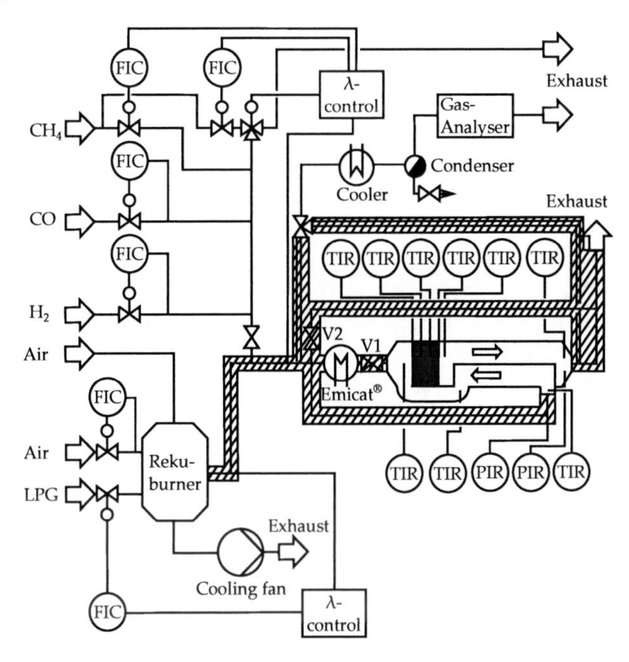

Figure A.3.: Rig for transient testing with near-stoichiometric conditions

A.2. Exhaust gas generator

For the experiments with near-stoichiometric exhaust gas composition (Section 5.2.4 and Section 5.3.3 (stationary) and Section 5.3.2 (transient)) an exhaust gas generator was applied. In a previous work, a fuel burner with similar specifications regarding maximum flow rate and exhaust composition was used for thermal aging of three-way catalyst bricks [10] ($T_{exh} > 1000°C$). Yet, in the present case the exhaust should be much cooler in order to prevent the heat exchangers from thermal damage and to specifically demonstrate the benefits regarding improved conversion performance with cold exhaust. Therefore, a recuperative gas burner was chosen which is able to generate exhaust with near-stoichiometric composition and low exhaust temperatures [78]. The main principle of this technology is flameless oxidation [76] which can be achieved by intensive mixing of feed gas with recirculated inert exhaust. As a result, formation of flames (i.e. hot-spots) is suppressed and the temperature distribution in the combustion chamber becomes very homogeneous leading to minimized NO_x emissions. The applied system (*WS REKUMAT C 150*) is depicted in Figure A.4.

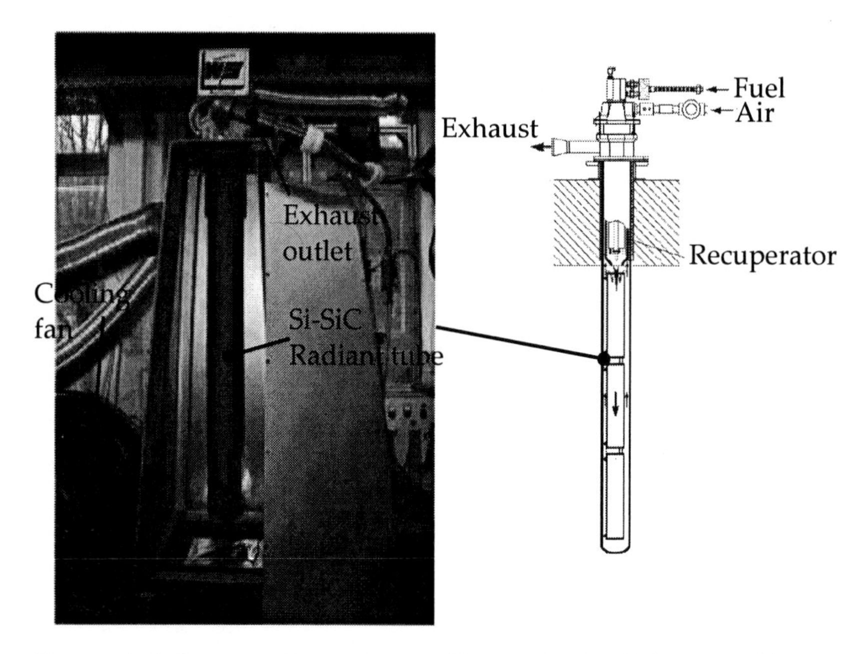

Figure A.4.: Left: recuperative gas burner with ceramic radiant tube mounted in support frame. Right: schematic of burner in radiant tube [77] with interior (i.e. recuperator, flame tubes, spacers).

The burner is attached to a ceramic radiant tube which emits the combustion heat to the surrounding cover plates of the support frame. For further cooling of the tube and mounting frame, a large fan ($\dot{V}_{air} = 50\ Nm^3/sec$) was installed. The cooling air flowed from bottom to top through the mounting frame.

Heat recuperation and the exhaust recirculation required for flameless oxidation are achieved by a special arrangement of burner and flame tubes shown schematically in Figure A.4, right. Between the burner's ceramic recuperator, the subjacently placed flame tubes and the surrounding radiant tube, an annular gap is formed. The cold feed is preheated by passing downwards through the interior of the recuperator. The hot exhaust passes in opposite direction through the adjacent gap. Between the end of the heat-exchanger zone and the top end of the flame tubes, the preheated feed stream is mixed with the exhaust passing through the annular gap in opposite direction. Hence, between the inside and outside of the flame tubes a circulating flow of hot exhaust emerges which leads to conditions suitable for flameless oxidation.

The applied *REKUMAT C 150* is specified for power outputs up to 40 kW [75]. However, in order to prevent the mounting frame from excessive temperatures during operation with increased power output, the system was usually operated below 20 kW. In Figure A.5, top, the resulting exhaust gas composition is shown for one operating point (6 kW, varying λ). The values obtained at higher flow rates were very similar and are therefore not shown. NO_x concentrations were measured but omitted in this plot since the concentrations were rather low ($< 100\,ppm$). In Figure A.5, bottom, the exhaust temperature is shown over the range of air flow rates usually used during heat exchanger testing. The obtained exhaust temperatures are well below the ones used during fuel-lean experiments. Yet, during cold start a considerable increase was obtained by the EMICAT system as shown in Section 5.3.2. During normal operation no additional heating was performed.

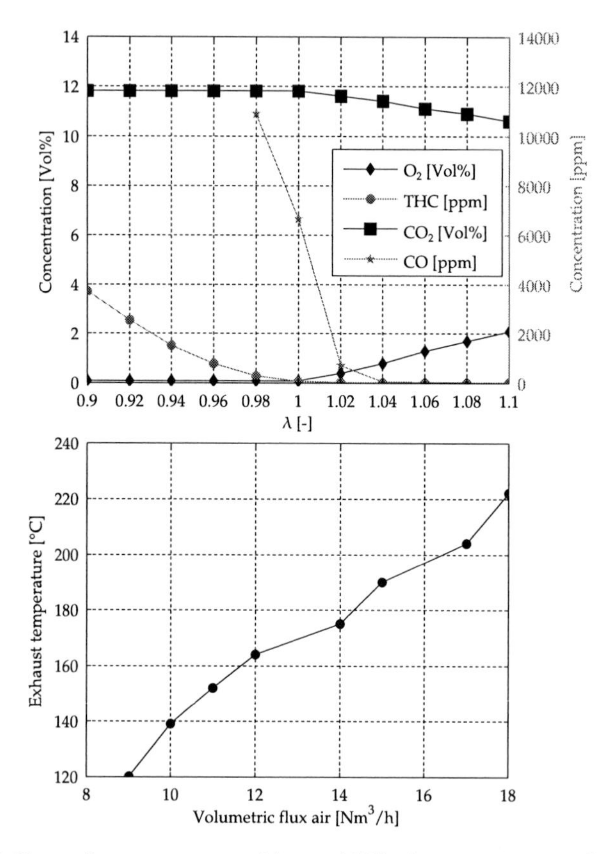

Figure A.5.: Top: exhaust gas composition at 6 kW of power output and varying λ. Bottom: Evolution of exhaust temperature over range of air flow rates for constant exhaust composition ($\lambda = 1.05$).

Appendix B.

Derivation of quasihomogeneous model equations

In this section, a detailed derivation of the quasihomogeneous heat-exchanger model is presented. It follows the principal ideas of the models developed in [23], [49] and [53]. In addition, two different reactions are assumed to occur on the catalyst surface. As demonstrated in Chapter 3, interesting effects can be observed if the ignition temperatures of these two reactions differs significantly. Initially, the quasihomogeneous model of a countercurrent reactor with two catalytic reactions occurring on the channel walls is derived. Subsequently, the same procedure is applied to the standard ceramic monolith case which is passed by the exhaust gas in one direction only. These models form the basis for all stationary simulation results presented in Chapter 3.

B.1. Catalytically coated, countercurrent reactor

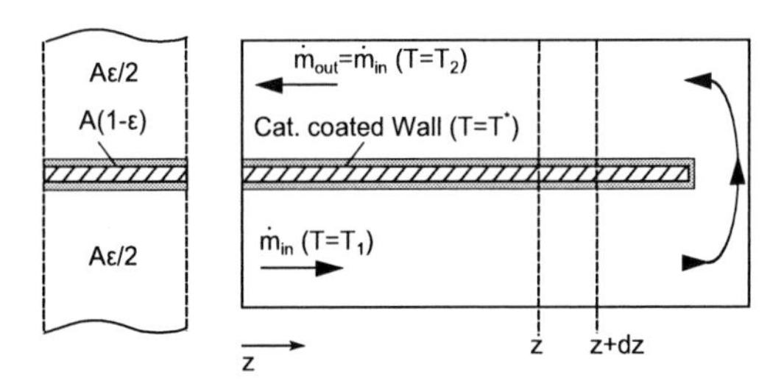

Figure B.1.: Simplified setup for quasihomogeneous steady-state model of countercurrent reactor

According to Figure B.1, in the fully coated prototype the gas enters the reactor at the left end, passes the catalyst coated channel wall and exits in opposite direction through the adjacent outflow channels. The gas void fraction ϵ and the specific exchange surface a_v are equally distributed over both channel sides. The specific mass flow $\dot{m}$ is referred to the total cross-sectional surface as in the detailed model. With these presumptions the stationary balance equations for the wall and the two gas phases read (see Equation 2.8 and Equation 2.1 for comparison):

Wall:

$$\begin{aligned} 0 &= (1-\epsilon)\cdot\lambda^s\cdot\frac{d^2T^*}{dz^2}+\alpha\cdot\frac{a_v}{2}\cdot(T_1-T^*)+\alpha\cdot\frac{a_v}{2}\cdot(T_2-T^*)+ \\ &+ \frac{a_v}{2}\cdot\sum_{i=1}^{I}(-\Delta h_{R,i})\cdot(r_{i,1}+r_{i,2}) \end{aligned} \tag{B.1}$$

Gas 1:

$$0 = -\dot{m}\cdot c_p^g\cdot\frac{dT_1}{dz}-\alpha\cdot\frac{a_v}{2}\cdot(T_1-T^*) \tag{B.2}$$

Gas 2:

$$0 = \dot{m}\cdot c_p^g\cdot\frac{dT_2}{dz}-\alpha\cdot\frac{a_v}{2}\cdot(T_2-T^*) \tag{B.3}$$

Addition of these three equations yields:

$$\begin{aligned} 0 &= (1-\epsilon)\cdot\lambda^s\cdot\frac{d^2T^*}{dz^2}+\frac{a_v}{2}\cdot\sum_{i=1}^{I}(-\Delta h_{R,i})\cdot(r_{i,1}+r_{i,2})+ \\ &+ \dot{m}\cdot c_p^g\cdot\left(\frac{dT_2}{dz}-\frac{dT_1}{dz}\right). \end{aligned} \tag{B.4}$$

Assuming axially constant physical properties and heat transfer coefficient and neglecting the dispersion terms, Equation B.2 and Equation B.3 can be derived with respect to z and subtracted from each other. This step yields the following equation:

$$\dot{m}\cdot c_p^g\cdot\left(\frac{d^2T_1}{dz^2}+\frac{d^2T_2}{dz^2}\right)=\alpha\cdot\frac{a_v}{2}\cdot\left(\frac{dT_2}{dz}-\frac{dT_1}{dz}\right) \tag{B.5}$$

For equal second derivatives (i.e. $\left(\frac{d^2T_{1,2}}{dz^2}\approx\frac{d^2T^*}{dz^2}\right)$, [74]) and symmetric enthalpy fluxes in both directions, the quasihomogeneous temperature can be approximated

as
$T^* = (T_1 + T_2)/2$. With this simplification and the substitution of equation B.5 into B.4 the final quasihomogeneous energy balance reads

$$0 = \lambda_{eff} \cdot \frac{d^2T^*}{dz^2} + \frac{a_v}{2} \cdot \sum_{i=1}^{I} (-\Delta h_{R,i}) \cdot (r_{i,1} + r_{i,2}), \tag{B.6}$$

with the effective heat conductivity as previously described in [37, 49, 53, 74]:

$$\lambda_{eff} = (1 - \epsilon) \cdot \lambda^s + \left(\dot{m} c_p^g\right)^2 \cdot \frac{4}{\alpha a_v}. \tag{B.7}$$

This quantity represents the impacts of axial heat conduction in the solid (first term) and the ratio of convective heat transport to heat transfer (second term) on the stationary axial temperature profile. In the next step, Equation B.7 is rewritten in order to show the physical background of the effective heat conductivity:

$$\lambda_{eff} = \dot{m} c_p^g L_{tot} \cdot \left(\underbrace{\frac{(1-\epsilon) \cdot \lambda^s}{\dot{m} \cdot c_p^g \cdot L_{tot}}}_{\widehat{=} Pe^{-1}} + \underbrace{\frac{4 \cdot \dot{m} \cdot c_p^g}{\alpha \cdot a_v \cdot L_{tot}}}_{\widehat{=} NTU^{-1}} \right). \tag{B.8}$$

The dimensionless ratio between convective heat transport and axial heat conduction in the solid corresponds to a Péclet number (Pe), the second term describing the dimensionless heat transfer has the shape of a number of transfer units (NTU). The specific mass flow can be eliminated by the space velocity τ with units $[1/s]$ using the following relation:

$$\tau := \frac{GHSV}{3600} = \frac{\dot{m}}{L_{tot} \rho^{N,in} \epsilon}, \tag{B.9}$$

with the gas hourly space velocity GHSV and the gas density $\rho^{N,in}$ at norm conditions (p = 1.013 bar, T = 273 K) and inflow gas composition. Hence, the following relations for Pe and NTU are obtained:

$$Pe = \frac{\tau \rho^{N,in} \epsilon c_p^g L_{tot}^2}{(1-\epsilon)\,\lambda^s}, \tag{B.10}$$

$$NTU = \frac{\alpha a_v}{4 \tau \rho^{N,in} \epsilon c_p^g} \tag{B.11}$$

In order to further simplify the quasihomogeneous energy balance, the second term of Equation B.6 needs to be modified, too. Therefore, in [23, 53] the reaction rates are eliminated using a simplified form of the material balance equations of gas and catalyst phase. This step is shown in the following for a system with two different reactions at each channel side. As in [23, 53], axial dispersion in the gas phases is neglected. Hence, the two gas phase material balances read:

$$0 = -\dot{m} \cdot \frac{dw_{j,1}^g}{dz} + \frac{a_v}{2} \cdot \beta \cdot \rho^g \cdot \left(w_{j,1}^s - w_{j,1}^g \right), \tag{B.12}$$

$$0 = \dot{m} \cdot \frac{dw_{j,2}^g}{dz} + \frac{a_v}{2} \cdot \beta \cdot \rho^g \cdot \left(w_{j,2}^s - w_{j,2}^g \right). \tag{B.13}$$

Accordingly, the solid phase material balance equations are:

$$MW_j \cdot \frac{a_v}{2} \cdot \sum_{i=1}^{I} \nu_{ij} r_{i,1} = \frac{a_v}{2} \cdot \beta \cdot \rho^g \cdot \left(w_{j,1}^s - w_{j,1}^g \right), \tag{B.14}$$

$$MW_j \cdot \frac{a_v}{2} \cdot \sum_{i=1}^{I} \nu_{ij} r_{i,2} = \frac{a_v}{2} \cdot \beta \cdot \rho^g \cdot \left(w_{j,2}^s - w_{j,2}^g \right). \tag{B.15}$$

The mass transfer term can be eliminated by insertion of B.14 and B.15 into B.12 and B.13, yielding:

$$0 = -\dot{m} \cdot \frac{dw_{j,1}^g}{dz} + MW_j \cdot \frac{a_v}{2} \cdot \sum_{i=1}^{I} \nu_{ij} r_{i,1}, \tag{B.16}$$

$$0 = \dot{m} \cdot \frac{dw_{j,2}^g}{dz} + MW_j \cdot \frac{a_v}{2} \cdot \sum_{i=1}^{I} \nu_{ij} r_{i,2}. \tag{B.17}$$

Addition of Equation B.16 and Equation B.17 and substitution into Equation B.6 yields:

$$0 = \dot{m}c_p^g L_{tot} \cdot \left(\frac{1}{Pe} + \frac{1}{NTU}\right) \cdot \frac{d^2T^*}{dz^2} + \dot{m} \cdot \sum_{i=1}^{I} \frac{-\Delta h_{R,i}}{MW_j \cdot \nu_{ij}} \cdot \left(\frac{dw_{j,1}}{dz} - \frac{dw_{j,2}}{dz}\right). \quad \text{(B.18)}$$

In this study, a system of two first-order reactions (i = 2) and two independent reactants $j = i = 2$ is considered. With this assumption and the introduction of the dimensionless coordinate $\zeta = z/L_{tot}$, Equation B.18 reads:

$$0 = \dot{m}c_p^g \cdot \left(\frac{1}{Pe} + \frac{1}{NTU}\right) \cdot \frac{d^2T^*}{d\zeta^2} + \dot{m} \cdot \sum_{i=1}^{I} \frac{-\Delta h_{R,i}}{MW_i} \cdot \left(\frac{dw_{i,2}}{d\zeta} - \frac{dw_{i,1}}{d\zeta}\right). \quad \text{(B.19)}$$

Now, the mass fractions w_i are replaced by the dimensionless conversion

$$\chi_j := \frac{w_{i,in} - w_i}{w_{i,in}}$$

and the equation can be further simplified:

$$0 = \left(\frac{1}{Pe} + \frac{1}{NTU}\right) \cdot \frac{d^2T^*}{d\zeta^2} + \sum_{i=1}^{I} \underbrace{\frac{w_{i,in} \cdot (-\Delta h_{R,i})}{MW_i \cdot c_p^g}}_{\Delta T_{ad,i}} \cdot \left(\frac{d\chi_{i,1}}{d\zeta} - \frac{d\chi_{i,2}}{d\zeta}\right). \quad \text{(B.20)}$$

If, as in case of this study, catalytic conversion of CO and CH_4 are considered as reactions, a further dimensionless parameter can be introduced, representing the distribution of the total adiabatic heat release $\Delta T_{ad,tot}$ over these two species:

$$\Psi := \Delta T_{ad,CO} / \Delta T_{ad,tot} \quad \text{(B.21)}$$

The final energy balance reads:

$$\begin{aligned} 0 \;=\; & \left(\frac{1}{Pe} + \frac{1}{NTU}\right) \cdot \frac{d^2T^*}{d\zeta^2} + \Delta T_{ad,tot} \cdot \left[\Psi \left(\frac{d\chi_{CO,1}}{d\zeta} - \frac{d\chi_{CO,2}}{d\zeta}\right) + \right. \quad \text{(B.22)} \\ + \; & \left. (1-\Psi)\left(\frac{d\chi_{CH_4,1}}{d\zeta} - \frac{d\chi_{CH_4,2}}{d\zeta}\right)\right]. \end{aligned}$$

Since this equation is of second order in space, two boundary conditions are required. The inflow condition derived in [53] was adapted to the present case, yielding:

$$T^*|_{\zeta=0} = T_{in} + \frac{\Delta T_{ad,tot}}{2} \cdot \left(\Psi \cdot \chi_{CO,2}|_{\zeta=0} + (1-\Psi) \cdot \chi_{CH_4,2}|_{\zeta=0}\right). \tag{B.23}$$

At the outflow end, a *Neumann* boundary condition is applied:

$$\left.\frac{dT^*}{d\zeta}\right|_{\zeta=1} = 0. \tag{B.24}$$

The material balances (i.e. Equation B.16 and Equation B.17) are transformed to the dimensionless length coordinate ζ and the space velocity τ:

$$0 = -\tau\rho^{N,in}\epsilon \cdot \frac{dw^g_{j,1}}{d\zeta} + MW_j \cdot \frac{a_v}{2} \cdot \sum_{i=1}^{I} \nu_{ij} r_{i,1}, \tag{B.25}$$

$$0 = \tau\rho^{N,in}\epsilon \cdot \frac{dw^g_{j,2}}{d\zeta} + MW_j \cdot \frac{a_v}{2} \cdot \sum_{i=1}^{I} \nu_{ij} r_{i,2}. \tag{B.26}$$

With the first-order rate expressions described in Section 2.1.1.7, these balance equations have units $[kg/m^3/s]$.

By the substitution $w_j = w_j^{in} - \chi_j w_j^{in}$, Equation B.25 and Equation B.26 can be expressed in terms of the conversion χ_j:

$$0 = \tau \cdot \rho^{N,in} \cdot \epsilon \cdot w_j^{in} \cdot \frac{d\chi_{j,1}}{d\zeta} + MW_j \cdot \frac{a_v}{2} \cdot \sum_{i=1}^{I} \nu_{ij} r_{i,1}, \tag{B.27}$$

$$0 = -\tau \cdot \rho^{N,in} \cdot \epsilon \cdot w_j^{in} \cdot \frac{d\chi_{j,2}}{d\zeta} + MW_j \cdot \frac{a_v}{2} \cdot \sum_{i=1}^{I} \nu_{ij} r_{i,2}, \tag{B.28}$$

Since dispersion effects are neglected, a *Dirichlet* boundary condition is assumed at the inflow end:

$$\chi_{j,1}|_{\zeta=0} = 0. \tag{B.29}$$

After leaving the inflow channels at the right end of the reactor, the gas flow is reversed and flows back to the left end through the outflow channels. At this transition point, ideal coupling is assumed:

$$\chi_{j,2}|_{\zeta=1} = \chi_{j,1}|_{\zeta=1}. \tag{B.30}$$

B.2. Standard ceramic monolith

In contrast to the previous case, the exhaust passes through the monolith channels in one direction only. An adapted schematic of this configuration is given in Figure B.2.

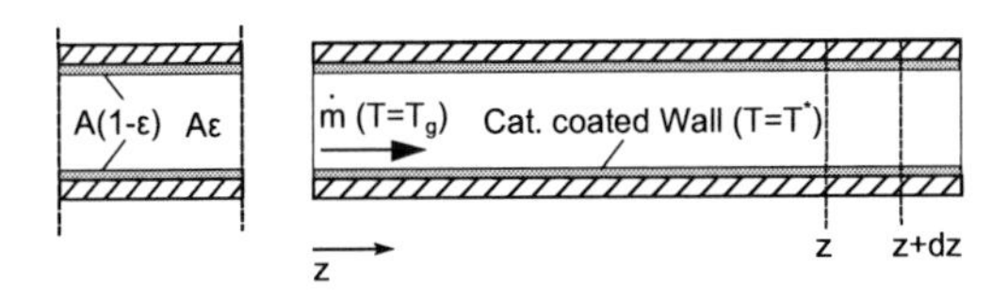

Figure B.2.: Simplified setup of standard ceramic monolith reactor

The stationary quasihomogeneous model for this setup was derived according to [49]. First, the energy balance equations for gas and wall phase are formulated:

Wall:

$$0 = (1-\epsilon)\,\lambda^{s}\frac{d^2T^*}{dz^2} + \alpha a_v\,(T^g - T^*) + a_v \cdot \sum_{i=1}^{I}(-\Delta h_{R,i}) \cdot (r_{i,1} + r_{i,2})\,, \quad \text{(B.31)}$$

Gas:

$$0 = -\dot{m}c_p^g\frac{dT^g}{dz} - \alpha a_v\,(T^g - T^*)\,. \quad \text{(B.32)}$$

Now, Equation B.31 and Equation B.32 are added and reformulated appropriately, yielding:

$$\begin{aligned} 0 &= \lambda^s\,(1-\epsilon)\,\frac{d^2T^*}{dz^2} - \dot{m}c_p^g\frac{(T^g - T^*)}{dz} - \dot{m}c_p^g\frac{dT^*}{dz} + \\ &+ a_v \cdot \sum_{i=1}^{I}(-\Delta h_{R,i}) \cdot (r_{i,1} + r_{i,2})\,. \end{aligned} \quad \text{(B.33)}$$

By differentiation of Equation B.32, the following relation is obtained:

$$-\frac{\dot{m}c_p^g}{\alpha a_v}\frac{d^2T^g}{dz^2} = \frac{(T^g - T^*)}{dz}. \quad \text{(B.34)}$$

With the assumption of [74], that

$$\frac{d^2T^g}{dz^2} \approx \frac{d^2T^*}{dz^2}, \tag{B.35}$$

and substitution of Equation B.34 into Equation B.33, the final quasihomogeneous energy balance is obtained:

$$0 = \lambda_{eff}\frac{d^2T^*}{dz^2} - \dot{m}c_p^g\frac{dT^*}{dz} + a_v \cdot \sum_{i=1}^{I}(-\Delta h_{R,i}) \cdot (r_{i,1} + r_{i,2}), \tag{B.36}$$

with

$$\lambda_{eff} = \lambda^s(1-\epsilon) + \left(\dot{m}c_p^g\right)^2 \cdot \frac{1}{\alpha a_v} = \dot{m}c_p^g L_{tot}\left(\underbrace{\frac{\lambda^s(1-\epsilon)}{\dot{m}c_p^g L_{tot}}}_{\widehat{=}Pe_{cat}^{-1}} + \underbrace{\frac{\dot{m}c_p^g}{\alpha a_v L_{tot}}}_{\widehat{=}NTU_{cat}^{-1}}\right). \tag{B.37}$$

Using Equation B.9, the dimensionless numbers are reformulated, yielding:

$$Pe_{cat} = \frac{\tau\rho^{N,in}\epsilon c_p^g L_{tot}^2}{(1-\epsilon)\lambda^s}, \tag{B.38}$$

$$NTU_{cat} = \frac{\alpha a_v}{\tau\rho^{N,in}\epsilon c_p^g} \tag{B.39}$$

Now, the dimensionless quasihomogeneous energy balance equation can be derived in direct analogy to Equation B.22.

$$\begin{aligned} 0 = & \left(\frac{1}{Pe_{cat}} + \frac{1}{NTU_{cat}}\right) \cdot \frac{d^2T^*}{d\zeta^2} - \frac{dT^*}{d\zeta} + \\ & + \Delta T_{ad,tot} \cdot \left[\Psi\frac{d\chi_{CO}}{d\zeta} + (1-\Psi)\frac{d\chi_{CH_4}}{d\zeta}\right]. \end{aligned} \tag{B.40}$$

Since in the standard ceramic monolith case the heat flux in the wall is not balanced by the two antipodal gas fluxes at the inflow end, the required boundary conditions are different to those derived for the countercurrent reactor. It is assumed, that at each end of the reactor heat transport to and from the system occurs exclusively by convection. In the system however, heat is also transported by axial conduction in the wall. As a result, at the inflow boundary the following balance equation has to be formulated:

$$T^*|_{z=0} = T_{in} + \frac{\lambda_{eff}}{\dot{m}c_p^g} \cdot \left.\frac{dT^*}{dz}\right|_{z=0}. \tag{B.41}$$

With the effective heat conductivity derived in Equation B.37 and the dimensionless length $\zeta = z/L_{tot}$ the final form of the boundary condition can be obtained:

$$T^*|_{\zeta=0} = T_{in} + \left(\frac{1}{Pe_{cat}} + \frac{1}{NTU_{cat}}\right) \cdot \left.\frac{dT^*}{d\zeta}\right|_{\zeta=0}. \tag{B.42}$$

At the right-hand side, an adiabatic condition is assumed:

$$\left.\frac{dT^*}{d\zeta}\right|_{\zeta=1} = 0. \tag{B.43}$$

The gas phase material balance with appropriate boundary conditions is equivalent to Equation B.25 and Equation B.29 which have been derived in the previous section.

Appendix C.

Approximation of light-off temperatures

In order to estimate the ignition temperature of the reactions occurring in the catalyst part of the derived reactor models, the theory of thermal explosion originally developed by [69] is applied. Based on this fundamental study a detailed derivation of the equations for an adiabatic batch reactor with zero-order exothermic reaction are given in [50]. In the present case, an equal procedure is applied for an adiabatic fixed-bed reactor, similar to the study performed by Zahn [79]. At first, an adiabatic energy balance of a fixed-bed reactor with zero-order reaction rate and no accumulation in the gas phase is formulated:

$$\epsilon \rho_g u_g c_{p,g} \frac{dT}{dz} = c_{A,0} \left(-\Delta h_R \right) k^0 e^{-\frac{E_a}{\mathbf{R}T}}, \tag{C.1}$$

with the gas velocity u_g, the gas void fraction ϵ, the heat capacity $c_{p,g}$, the initial concentration of the reactant c_A^0 and kinetic parameters. With the adiabatic temperature rise

$$\Delta T_{ad} = \frac{c_{A,0} \left(-\Delta h_R \right)}{\rho_g c_{p,g}} \tag{C.2}$$

substituted into Equation C.1 the energy balance reads:

$$u_g \frac{dT}{dz} = \Delta T_{ad} k^0 e^{-\frac{E_a}{\mathbf{R}T}}. \tag{C.3}$$

Now, the Arrhenius term can be reformulated introducing the dimensionless temperature $\Theta := \frac{T - T_0}{\Delta T_{ad}}$:

$$k^0 e^{-\frac{E_a}{\mathbf{R}T}} = k\left(T_0\right) e^{\frac{E_a}{\mathbf{R}} \left(\frac{1}{T_0} - \frac{1}{T} \right)} = k\left(T_0\right) e^{\frac{E_a \Delta T_{ad}}{\mathbf{R}T T_0} \Theta}. \tag{C.4}$$

Equation C.4 can be further simplified, assuming that $T \approx T_0$:

$$k\left(T_0\right) e^{\frac{E_a \Delta T_{ad}}{\mathbf{R} T T_0}\Theta} \leq k\left(T_0\right) e^{\overbrace{\frac{E_a \Delta T_{ad}}{\mathbf{R} T_0^2}}^{:=\gamma}\Theta}, \tag{C.5}$$

with the dimensionless activation energy γ. With the Arrhenius term (Equation C.5) substituted into the original energy balance (Equation C.3) and transformation of T into Θ, the following relation is obtained:

$$u_g \frac{d\Theta}{dz} \leq k\left(T_0\right) e^{\gamma\Theta}. \tag{C.6}$$

By introduction of the dimensionless length $\zeta = z/L_{tot}$ Equation C.6 can be further modified, yielding:

$$\frac{1}{\tau}\frac{d\Theta}{d\zeta} \leq k\left(T_0\right) e^{\gamma\Theta}, \tag{C.7}$$

with the residence time $\tau = L_{tot}/u_g$.

C.1. Ignition time

With Equation C.7 a relation between the time span required for full pollutant conversion and liberation of the adiabatic temperature rise and residence time in the reactor is given. In fact, this critical time is equal to the ignition time of the reaction. After separation of variables, Equation C.7 can be integrated once:

$$\frac{1}{\tau}\int_0^1 \frac{d\Theta}{e^{\gamma\Theta}} \leq \int_0^1 k\left(T_0\right) d\zeta, \tag{C.8}$$

yielding

$$\tau \geq \frac{1 - \overbrace{e^{-\gamma}}^{\ll 1}}{\gamma k\left(T_0\right)} \approx \frac{1}{\gamma k\left(T_0\right)}. \tag{C.9}$$

The exponential term in Equation C.9 can be usually neglected. Now, the final balance equation can be written after insertion of the original definitions of τ and γ:

$$\frac{L_{tot}}{u_g} \geq \frac{1}{k^0 e^{-\frac{E_a}{\mathbf{R}T_0}}} \cdot \frac{\mathbf{R}T_0^2}{\Delta T_{ad} E_a}. \tag{C.10}$$

C.2. Estimation of ignition temperatures in case of multiple reactions

Equation C.10 shows the final relation between residence time τ and the inflow temperature T_0 required for full conversion of the pollutant. Hence, in the following this special inflow temperature for a given τ will be denoted as "ignition temperature" T_{ign}. In the present case, two reactions occur in the fixed-bed reactor. First, CO is converted liberating a defined portion of the adiabatic temperature rise. After ignition of the second component, CH_4, the complete adiabatic temperature rise $\Delta T_{ad,tot}$ is obtained. Therefore, two different dimensionless activation energies have to be formulated based on the parameters introduced in Section 2.2:

$$\frac{1}{\gamma_{CO}} = \frac{1}{\Psi \cdot \Delta T_{ad,tot}} \cdot \frac{\mathbf{R}T_0^2}{E_{A,CO}} \tag{C.11}$$

$$\frac{1}{\gamma_{CH_4}} = \frac{1}{\Delta T_{ad,tot}} \cdot \frac{\mathbf{R}T_0^2}{E_{A,CH_4}}. \tag{C.12}$$

These two parameters can now be inserted into Equation C.10, whose units have to be adapted to comply with the units of the kinetic parameters used in this study, yielding the final relation for the two ignition temperatures:

$$\frac{p_{gas} MW_{gas}}{\mathbf{R}T_{ign,j} \rho^N \tau} = \frac{\rho_0 \epsilon}{a_v MW_{gas} k_{0,j} e^{-\frac{E_{A,j}}{\mathbf{R}T_{ign,j}}}} \cdot \frac{1}{\gamma_j}. \tag{C.13}$$

In conclusion, the derived equation allows to estimate the ignition behavior of exothermic reactions in a fixed-bed reactor based on the geometric and kinetic properties of the system. In Section 3.2.1 it is shown that the results are similar to the usually obtained T_{50} light-off temperature.

Appendix D.

Geometric and thermophysical properties of simulation models

D.1. Geometric properties of different reactor prototypes

D.1.1. Folded-sheet prototype

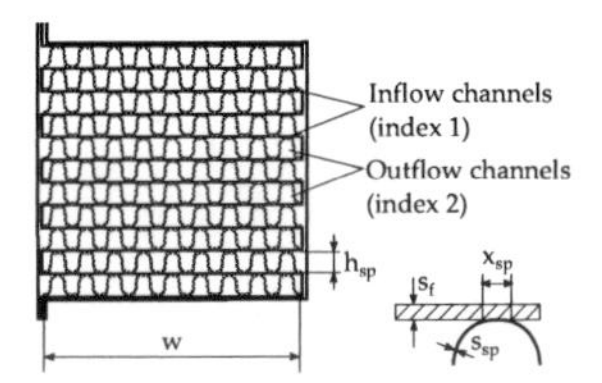

Figure D.1.: Geometry of folded-sheet heat exchanger with corrugated spacer structures (insulation omitted)

Calculation of geometric properties for this prototype was directly adopted from [29] with slight modifications. In Figure D.1 the geometry of the folded-sheet heat exchanger is schematically depicted. The channel height is determined by the height of the spacer structures h_{sp} whose shape is assumed to be semi-circular. Certain values of metal sheet thickness are assumed for the spacers (s_{sp}) and the wall material s_f. The corrugated spacer structures are in close contact with the wall over a certain length x_{sp} where contact heat transfer occurs. This value was estimated to 0.3 mm for the laboratory-scale prototype depicted in Figure 5.2. Contrary to [29], only one layer of spacer structures per channel is considered. By adding the height of the single channels and multiplying with the total width, the total cross-sectional area can be calculated as:

$$A_{tot} = \left((n_{ch,1} + n_{ch,2}) \cdot h_{sp} + (n_{ch,1} + n_{ch,2}) \cdot s_f + 2 \cdot s_f \right) \cdot (w + 3 \cdot s_f), \qquad \text{(D.1)}$$

with the number of inflow channels $n_{ch,1}$ and the number of outflow channels $n_{ch,2}$. According to the construction, both the first and the last seam of the folded-sheet structure are located at the left-hand side and tightly welded to the lateral plate with the sidewise outlet (see Figure 5.2 and Figure D.1). As a consequence, the number of outflow channels exceeds the number of inflow channels by one.

In the next step, the volume fractions of the five different phases are calculated. Unlike the assumption in [29], that the casing is directly welded to the folded-sheet structure and the insulation is applied externally, in the present case tight fixation of the heat exchanger core was achieved by a compression jacket made of the same thin material as the folded sheet structure. Around this inner core, $\sim$ 20 mm of ceramic insulation were applied before the outer canning was attached. In Figure D.2, a schematic of this layout is shown. In this study, only the parts below the internal insulation jacket are considered with the outer casing neglected.

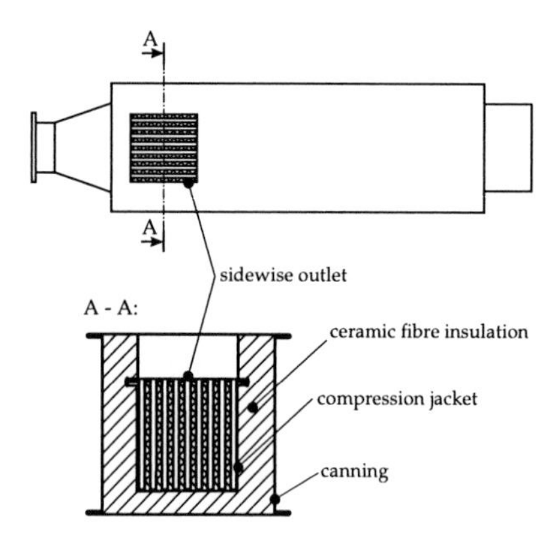

Figure D.2.: Insulated and canned heat exchanger (top). Sectional view of sidewise outlet (bottom).

These assumptions lead to the following two relations for the void fraction of the folded-sheet wall ϵ_{fs} and the compression jacket ϵ_{cs}. These two volume fractions are treated as one homogeneous phase ϵ_s:

$$\epsilon_f = \frac{\left((n_{ch,1}+n_{ch,2})\cdot h_{sp} + (n_{ch,1}+n_{ch,2})\cdot(w+s_f)\right)\cdot s_f}{A_{tot}} \quad \text{(D.2)}$$

$$\epsilon_{cs} = \frac{2\cdot\left((n_{ch,1}+n_{ch,2})\cdot(h_{sp}+s_f)+2\cdot s_f+w\right)\cdot s_f}{A_{tot}} \quad \text{(D.3)}$$

$$\epsilon_s = \epsilon_{fs}+\epsilon_{cs}. \quad \text{(D.4)}$$

Similar equations can be formulated for the volume fractions of the corrugated spacers ϵ_k^c and the gas phases ϵ_k^g at either channel side k:

$$\epsilon_k^c = \frac{0.5\cdot\pi\cdot h_{sp}\cdot s_{sp}\cdot\frac{w}{h_{sp}}\cdot n_{ch,k}}{A_{tot}}, \qquad k=1,2 \quad \text{(D.5)}$$

and :

$$\epsilon_k^g = \frac{h_{sp}\cdot n_{ch,k}\cdot w}{A_{tot}} - \epsilon_k^c, \qquad k=1,2. \quad \text{(D.6)}$$

The specific surfaces for exchange of heat and mass a_v with units $\left[\frac{m^2}{m_{tot}^3}\right]$ are formulated according to the following relations:

Exterior surface of heat exchanger core for heat losses:

$$a_v^{s,amb} = \frac{2\cdot\left((n_{ch,1}+n_{ch,2})\cdot h_{sp}+(n_{ch,1}+n_{ch,2}+1)\cdot s_f+3\cdot s_f+w\right)}{A_{tot}}. \quad \text{(D.7)}$$

Contact surface between spacers and wall at either channel side k:

$$a_{v,k}^{c,s} = \frac{\frac{w}{h_{sp}}\cdot n_{ch,k}\cdot x_{sp}}{A_{tot}}, \qquad k=1,2. \quad \text{(D.8)}$$

Exchange surface between gas and spacer structure at either channel side k:

$$a_{v,k}^{g,c} = \frac{\pi\cdot h_{sp}\cdot\frac{w}{h_{sp}}\cdot n_{ch,k}}{A_{tot}} - a_{v,k}^{c,s}, \qquad k=1,2. \quad \text{(D.9)}$$

Exchange surface between gas and folded-sheet wall at either channel side k:

$$a_{v,k}^{g,s} = \frac{2\cdot(w+h_{sp,k})\cdot n_{ch,k}}{A_{tot}} - a_{v,k}^{c,s}, \qquad k=1,2. \quad \text{(D.10)}$$

It is important to note that the sidewise outlet is neglected in the calculation of the outflow channel's exchange surface.

D.1.2. Brazed prototype

As shown in Section 5.2, in the brazed prototype a stack of flat tubes with interjacent corrugated spacers is brazed in a high-temperature vacuum process. Inflow and outflow channels are separated by a header plate which closes the outflow channels axially at the inflow end of the heat exchanger. Due to the tight brazing, no compression jacket is required and the heat exchanger can be directly placed in an insulated casing. The resulting geometric layout as assumed for the simulations is depicted in Figure D.3.

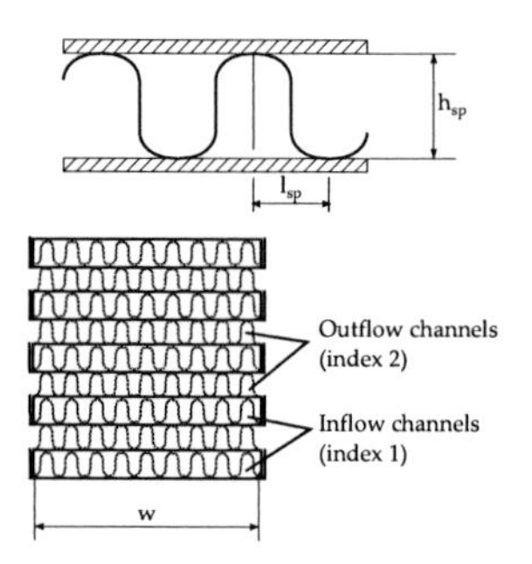

Figure D.3.: Geometry of brazed heat exchanger with corrugated spacer structures (insulation omitted)

First, the relation for the total cross-sectional area is formulated:

$$A_{tot} = \left(\left(n_{ch,1} + n_{ch,2}\right) \cdot h_{sp} + 2 \cdot n_{ch,1} \cdot s_f\right) \cdot \left(w + 4s_f\right). \tag{D.11}$$

As a consequence of the direct brazing without compression jacket, only one volume fraction ϵ_s for the walls of the flat tubes is needed:

$$\epsilon_s = \frac{\left(4 \cdot n_{ch,1} \cdot h_{sp} + 2 \cdot n_{ch,1} \cdot \left(w + 4 \cdot s_f\right)\right) \cdot s_f}{A_{tot}}. \tag{D.12}$$

The spacer structures are different from the ones applied in the folded-sheet prototype. There, a sequence of semi-circular elements was assumed as approximation of the spacer profiles. In the brazed prototype however, the spacer height h_{sp} is usually grater than the width of one element l_{sp}. The relations for the volume fractions of the spacers ϵ_k^c and of the gas phases ϵ_k^g read:

$$\epsilon_k^c = \frac{\left(0.5 \cdot \pi \cdot l_{sp} + \left(h_{sp} - l_{sp}\right)\right) \cdot s_{sp} \cdot \frac{w}{l_{sp}} \cdot n_{ch,k}}{A_{tot}}, \qquad k = 1,2 \tag{D.13}$$

and :

$$\epsilon_k^g = \frac{h_{sp} \cdot n_{ch,k} \cdot w}{A_{tot}} - \epsilon_k^c, \qquad k = 1,2. \tag{D.14}$$

As for the folded-sheet prototype the different exchange surfaces with units $\frac{m^2}{m^3_{tot}}$ are formulated starting with the exterior surface for heat losses $a_v^{s,amb}$:

$$a_v^{s,amb} = \frac{2 \cdot \left(\left(n_{ch,1} + n_{ch,2}\right) \cdot h_{sp} + 2 \cdot n_{ch,1} \cdot s_f + 4 \cdot s_f + w\right)}{A_{tot}}. \tag{D.15}$$

Since the spacer structures were brazed to the walls and wider radii were applied for the corrugation, the contact width x_{sp} was increased to 0.5 mm for the brazed prototype. The relation for the specific contact surface in either channel reads:

$$a_{v,k}^{c,s} = \frac{\frac{w}{l_{sp}} \cdot n_{ch,k} \cdot x_{sp}}{A_{tot}}, \qquad k = 1,2. \tag{D.16}$$

The relation for the specific surface between gas and spacer structures is formulated accordingly:

$$a_{v,k}^{g,c} = \frac{\left(\pi \cdot l_{sp} + 2 \cdot \left(h_{sp} - l_{sp}\right)\right) \cdot \frac{w}{l_{sp}} \cdot n_{ch,k}}{A_{tot}} - a_{v,k}^{c,s}, \qquad k = 1,2. \tag{D.17}$$

For the specific surface between gas and wall it is important to note that only the inflow channels (i.e. index 1) are completely surrounded by wall material. In the outflow channels the lateral sealing is achieved by tight brazing of the corrugated spacers. Consequently, the relation for $a_{v,1}^{g,s}$ and $a_{v,2}^{g,s}$ read:

$$a_{v,1}^{g,s} = \frac{2 \cdot \left(w \cdot n_{ch,1} + h_{sp,1} \cdot n_{ch,1}\right)}{A_{tot}} - a_{v,1}^{c,s}, \tag{D.18}$$

and :

$$a_{v,2}^{g,s} = \frac{2 \cdot w \cdot n_{ch,2}}{A_{tot}} - a_{v,2}^{c,s}. \tag{D.19}$$

D.1.3. Ceramic monolith

For the standard system without heat exchanger and the sequential heat-integrated case, a coated ceramic honeycomb with quadratic channels is assumed as catalyst substrate. In Figure D.4 its geometric layout is schematically depicted. For the simu-

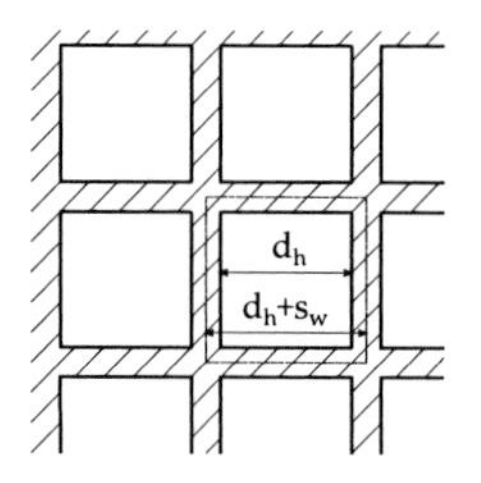

Figure D.4.: Geometric layout of standard ceramic monolith

lations performed, the catalyst washcoat is assumed to be part of the channel walls. Hence, the wall thickness s_w is assumed to comprise both the honeycomb base material and the washcoat. With the cell density d_c and the wall thickness s_w given, the required geometric parameters can be calculated. The hydraulic diameter d_h directly follows as:

$$d_h = \sqrt{\frac{1}{d_c}} - s_w. \tag{D.20}$$

With this value obtained, the remaining geometric parameters a_v (specific exchange surface for mass and heat transfer) and ϵ_g (void fraction) can be calculated:

$$a_v = \frac{4d_h}{(d_h + s_w)^2}, \tag{D.21}$$

$$\epsilon_g = \frac{d_h^2}{(d_h + s_w)^2}. \tag{D.22}$$

The exterior specific surface $a_v^{s,amb}$ over which heat losses occur is calculated as ratio between outer surface of the cylindric monolith brick to its total volume:

$$a_v^{s,amb} = \frac{\pi D L_{tot}}{\frac{\pi}{4} D^2 Ltot} = \frac{4}{D}, \tag{D.23}$$

with the total diameter of the ceramic honeycomb D. For a brick with rectangular cross section, as assumed for simulation of the sequential system, the relation has to be formulated differently:

$$a_v^{s,amb} = \frac{2 \cdot (w + h)}{wh}, \tag{D.24}$$

with the brick's width w and height h.

D.2. Properties of gas phases and solid materials

D.2.1. Gas density

Both gas phases are assumed to be ideal due to the moderate pressure level. Hence, the density of the respective gas phase k is calculated as:

$$\rho_k^g = \frac{p}{\mathbf{R}T_k^g} \cdot \frac{1}{\sum_j \frac{w_j^g}{MW_j}}, \tag{D.25}$$

with the mass fractions w_j^g and molecular weights MW_j of the different gas phase components.

D.2.2. Heat capacity and enthalpy

Due to the generally low pressure level, both the heat capacities and the enthalpies are assumed to depend only on temperature. For the specific heat capacities of the different gas phase species with units $\left[\frac{J}{molK}\right]$, the following polynomial approach is applied [41]:

$$c_{p,j}\left(T\left[K\right]\right) = a_j + b_j \cdot T + c_j \cdot T^2 + d_j \cdot T^3 + e_j \cdot \frac{1}{T^2}. \tag{D.26}$$

The required fitting parameters are listed in Table D.1. In order to obtain the total mass-related heat capacity of both gas phases, the component's heat capacities are converted and summed:

$$c_p^g\left(T, w_j\right) = \sum_j \frac{w_j}{MW_j} \cdot c_{p,j}. \tag{D.27}$$

For calculation of the specific enthalpies Equation D.26 is integrated once:

$$h_j\left(T\left[K\right]\right) = h_j^0 + a_j \cdot T + \frac{b_j}{2} \cdot T^2 + \frac{c_j}{3} \cdot T^3 + \frac{d_j}{4} \cdot T^4 - \frac{e_j}{T}. \tag{D.28}$$

The parameters required for evaluation of Equation D.26 and Equation D.28 are listed below:

Table D.1.: Parameters for calculation of heat capacities and enthalpies of different gas phase species [43, 41].

	a_j	$b_j \cdot 10^{-3}$	$c_j \cdot 10^{-6}$	$d_j \cdot 10^{-9}$	$e_j \cdot 10^{6}$	$h_j^0 \left[\frac{J}{mol}\right]$
N_2	19.506	19.887	-8.599	1.370	0.527	0.0
O_2	30.032	8.773	-3.988	0.788	-0.741	0.0
CH_4	-0.703	108.477	-42.522	5.863	0.678	-74873
CO	25.568	6.096	4.055	-2.671	0.131	-110527
CO_2	24.997	55.187	-33.691	7.948	-0.137	-393522
H_2O	30.092	6.833	6.793	-2.534	0.082	-241826

Reaction enthalpies for the catalytic conversion of CO and CH_4 are calculated with:

$$\Delta h_{R,i}(T) = \sum_j \nu_{i,j} \cdot h_j(T). \tag{D.29}$$

D.2.3. Viscosity

The gas phases' dynamic viscosity η with units $\left[\frac{kg}{ms}\right]$ is calculated according to relations given in [63] for ideal mixtures:

$$\eta = \sum_i \frac{y_i \eta_i(T)}{\sum_j y_j \Phi_{i,j}}, \tag{D.30}$$

with $\Phi_{i,j}$ approximated as $\sqrt{MW_j/MW_i}$. As further simplification, only interaction with N_2 is considered (i.e. $j = N_2 = const.$) since this is the predominant gas component. Hence, the following expression is obtained for the viscosity of the gas mixture:

$$\eta = \sum_i \frac{y_i \sqrt{MW_i}}{y_{N_2} \sqrt{MW_{N_2}}} \cdot \eta_i(T). \tag{D.31}$$

The components' viscosity is calculated based on a polynomial approximation with parameters taken from [27]:

$$\eta_i(T[K]) = a_i + b_i \cdot T + c_i \cdot T^2 + d_i \cdot T^3 + e_i \cdot T^4. \tag{D.32}$$

Table D.2.: Parameters for calculation of dynamic viscosity η_i with units μPas [27].

	$a_i \cdot 10^{-5}$	$b_i \cdot 10^{-7}$	$c_i \cdot 10^{-10}$	$d_i \cdot 10^{-12}$	$e_i \cdot 10^{-15}$
N_2	-0.01020	0.74785	-0.59037	0.03230	-0.00673
O_2	-0.10257	0.92625	-0.80657	0.05113	-0.01295
CH_4	-0.07759	0.50484	-0.43101	0.03118	-0.00981
CO	0.01384	0.74306	-0.62996	0.03948	-0.01032
CO_2	-0.18024	0.65989	-0.37108	0.01586	-0.00300
H_2O	-0.10718	0.35248	0.03575	-	-

D.2.4. Heat conductivity

In complete analogy to dynamic viscosity, the gas phases' heat conductivity λ with units $\left[\frac{W}{mK}\right]$ is calculated according to [63]:

$$\lambda = \sum_i \frac{y_i \lambda_i}{\sum_j y_j A_{i,j}}. \tag{D.33}$$

With the assumption from the previous section, that N_2 is the main constituent of both gas phases, and the analogy $A_{i,j} = \Phi_{i,j} = \sqrt{MW_j / MW_i}$, the relation for the gas phases' heat conductivity is yielded:

$$\lambda = \sum_i \frac{y_i \sqrt{MW_i}}{y_{N2} \sqrt{MW_{N_2}}} \cdot \lambda_i. \tag{D.34}$$

As for the dynamic viscosity, a polynomial approach is applied to calculate the components' heat conductivity at the respective temperature [27]:

$$\lambda_i\left(T\left[K\right]\right) = a_i + b_i \cdot T + c_i \cdot T^2 + d_i \cdot T^3 + e_i \cdot T^4, \tag{D.35}$$

with the required parameters shown in Table D.3.

Table D.3.: Parameters for calculation of gas phase heat conductivity λ_i with units $W/m/K$ [27].

	$a_i \cdot 10^{-3}$	$b_i \cdot 10^{-3}$	$c_i \cdot 10^{-6}$	$d_i \cdot 10^{-9}$	$e_i \cdot 10^{-12}$
N_2	-0.13	0.101	-0.060650	0.033610	-0.007100
O_2	-1.29	0.107	-0.052630	0.025680	-0.005040
CH_4	8.15	0.008	0.351530	-0.338650	0.140920
CO	-0.78	0.103	-0.067590	0.039450	-0.009470
CO_2	-3.882	0.053	0.071460	-0.070310	0.018090
H_2O	0.46	0.046	0.051150	-	-

D.2.5. Diffusion coefficients

The diffusion coefficient $D_{j,k}$ of species j in a homogeneous mixture k was adopted from [63]:

$$D_{j,k} = \left(\sum_{\substack{i=1 \\ i \neq j}}^{n} \frac{y_i}{D_{j,i}} \right)^{-1}. \tag{D.36}$$

For this formula, the binary diffusion coefficients $D_{j,i}$ with units $\left[\frac{m^2}{s}\right]$ are required which were obtained by the correlation according to *Fuller, Schettler* and *Giddings* [63]:

$$D_{j,i}\left(T\,[K], p\,[bar]\right) = \frac{1.43 \cdot 10^{-7} T^{1.75} \sqrt{\frac{MW_j + MW_i}{MW_j \cdot MW_i}}}{\sqrt{2} p \left[\left(\sum \nu_j\right)^{1/3} + \left(\sum \nu_i\right)^{1/3} \right]^2}. \tag{D.37}$$

The required diffusion volumina $\nu_{i/j}$ are listed in Table D.4. As previously assumed

Table D.4.: Diffusion volumina of gas phase components [63].

N_2	O_2	CH_4	CO	CO_2	H_2O
18.5	16.3	25.14	18.0	26.9	13.1

for the binary interaction parameters in the relations for dynamic viscosity and heat conductivity, only interaction bewteen N_2 and the remaining gas species is considered.

D.2.6. Properties of reactor materials

The material parameters for the metallic heat exchangers and the ceramic honeycomb substrates were adopted from [29] and [57]. The applied values are listed in Table D.5. The temperature dependence was neglected.

Table D.5.: Thermophysical properties of solid phases. Values for stainless steel were adopted from [29], those for cordierite from [57].

	$\rho^s \left[\frac{kg}{m^3}\right]$	$c_p^s \left[\frac{kJ}{kgK}\right]$	$\lambda^s \left[\frac{W}{mK}\right]$
Stainless steel	7800	0.58	22
Cordierite	1632	1.05	2

D.3. Transport parameters

D.3.1. Axial dispersion of heat and mass

In accordance with [29], axial dispersion was implemented by averageing gas phase composition and temperature at heat exchanger inlet for the usually applied conditions. Yet, due to the clear domination of convection, the influence of axial dispersion is rather small. Eventually, the same set of parameters (i.e. $D_{ax}^g \approx 0.002\ m^2/s$ and $\lambda_{ax}^g \approx 2\ W/m/K$) as in [29] was applied.

D.3.2. Coefficients for heat and mass transfer

D.3.2.1. Heat transport

Two different cases of heat transport are considered. In the first one, heat exchange occurs between the two gas phases and channel walls or spacer surfaces. In addition to that, heat is exchanged over the contact surfaces between spacer structures and the channel walls. For the first case, a relation for fully developed laminar flow was applied [1]:

$$\alpha_k^g(T) = Nu\frac{\lambda_k^g(T)}{d_h}. \tag{D.38}$$

The assumptions of the channels as circular tubes and constant heat flux density at the solid surfaces lead to a constant average Nusselt number of 3.74. The heat conductivity λ is calculated with Equation D.34. The hydraulic diameter is calculated for the different reactor geometries. The relation for the ceramic honeycomb with rectangular channels was already mentioned (Equation D.20). For the folded-sheet prototype, the relation is derived based on the geometric layout shown in Figure D.1 according to [29]:

$$d_h = \frac{4 \cdot A}{U} = \frac{4 \cdot \epsilon_k^g \cdot A_{tot}}{\left(\pi + 2 - \frac{x_{sp}}{h_{sp}}\right) \cdot w \cdot n_{ch,k}}. \tag{D.39}$$

Along similar lines, an expression was derived based on Figure D.3 for the brazed prototype:

$$d_h = \frac{4 \cdot A}{U} = \frac{4 \cdot \epsilon_k^g \cdot A_{tot}}{\left(\pi \cdot l_{sp} + 2 \cdot h_{sp} - x_{sp}\right) \cdot \frac{w}{l_{sp}} \cdot n_{ch,k}}. \tag{D.40}$$

The influence of contact heat transfer between corrugated spacer structures and wall on reactor performance was discussed in [29]. In that work, the catalyst was only applied on the spacer structures. Hence, heat exchange between spacers and wall was required in order to avoid excessive temperatures of the catalyst phase. In the present case however, the catalyst is applied on both the walls and the spacers. Moreover, in case of the brazed prototype the structures are tightly connected to the adjacent walls. As a result, the temperature difference between spacers and wall is expected to be rather small. Therefore, a very high contact transfer coefficient ($\alpha_k^{c,s} = 10\, kW/m^2/K$) was applied which virtually merges spacers and wall.

For the heat losses through the heat exchangers' insulation jackets to ambience, a constant value of $\alpha^{s,amb} = 1.5\, W/m^2/K$ was assumed initially. This value was slightly modified during model adaption (see Sections 5.1.2.1 to 5.2.2). Generally, the results with one constant parameter agreed quite well with the experimental results which is why a temperature-dependent approach, as proposed in [29], was not pursued.

D.3.2.2. Mass transport

The mass transfer coefficients of the different components directly follow from the definition of the Sherwood number [1]:

$$\beta_{j,k} = \frac{D_{j,k} \cdot Sh}{d_h}. \tag{D.41}$$

The hydraulic diameter is calculated with the relations derived in the previous section and the diffusion coefficient of component j in gas phase k follows from Equation D.36 with N_2 as the only interacting species. Furthermore, assuming direct analogy between heat and mass transfer (i.e. $Nu \approx Sh$) and constant concentration of the respective gas component at the catalyst surface, a Nusselt number of 3.07 is obtained [1]. As simplification, the mass transfer coefficient is set equal for all components considered in this work. Both, [29] and [23] have discussed the effect of this step and come to the conclusion, that the error caused by this simplification is rather small.